Tommy Melton
NE4TM
2000

The COMPLETE DX'ER

Second Edition
by Bob Locher, W9KNI
with drawings by Wayne T. Pierce, K3SUK

Published by Idiom Press,
P. O. Box 583
Deerfield, Illinois 60015
U. S. A.
Copyright 1989

ISBN 0-9617577-0-1

Dedication

First, to my wife, Judy, who graciously puts up with me and DX'ing.
Second, to my parents, who supported my early days as a DX'er.
Third, to the memory of Art Collins, WØCXX, who got me into amateur radio, and whose shadow will always lie long over high frequency amateur radio operation.
Fourth, to Bob White, whose devoted stewardship of the DXCC program made DX'ing the wonderful game that it is today.
And last, to the memory of Larry LeKashman, W2IOP, W9IOP, W2AB, whose book, "CQ DX" was the first real book on DX'ing, and who was a good friend.

Thanks also to the many people without whose help this book would never have been written, including WA9SLD, W9NB, K9MM, W9NUD, W9WU, and others. Special thanks are due to Jim Rafferty, N6RJ, who served as advisor, to Herb Nelson, W9IGL, whose eagle eye as line editor is unmatched, to Gene Gauss, who is responsible for the design, and to Jere Benedict, my business partner in the real world, for whose help, assistance and patience I cannot thank enough.

This book, now in a fully updated second edition, is written for the radio amateur who is fascinated by DX and the working of foreign countries, and who wishes to be successful at it. If you can work both coasts, you can work DX. This work is intended to help show you the way, from your first trans-oceanic QSO to the top of the DXCC Honor Roll.

We make the assumption that the goal of the DX'er is to work a new one. To achieve this goal, the DX'er always should be trying to optimize the odds of success. In most cases the easiest way to work a new country is to find a station from that country calling CQ, or signing clear from a QSO without anyone else being aware that he is there, so that the DX'er gets into a one-on-one situation. To do this regularly and successfully means to listen, listen, listen. This is the first operating technique of the DX'er that will be covered in detail.

Another way to log a new one is to have him respond to your CQ. Calling CQ DX is widely looked down upon as the least efficient way of all to chase DX, and has been vilified for years. Yet, the big multi-multi contest stations will work over 100 countries in one weekend on each of three or four bands, with country counts of 165 countries or more in one weekend not unheard of. And most of those countries were worked in response to CQ calls by the contest station. The technique of the CQ DX call will carefully examined. It can be a useful tactic.

Speaking of contests, they offer a terrific opportunity to run up a country total in a hurry. A really good single operator with a good station and good conditions can often work 150 countries in a single weekend in the CQ WW Contest, especially if he is purely out for country totals and has no intention of optimizing a contest score.

The author worked 89 countries in one contest on fifteen meters only in one weekend, and 96 different countries in a multi-band effort the following year under indifferent propagation conditions. While this is not by some standards an impressive effort, still, the author did it in the QRP class with

III

five watts maximum output at all times. Contests as vehicles for country hunting will be considered at greater length.

Frequently a new country is discovered by the pile-up of stations already in hot pursuit. This is a very easy way to find new countries, but working them can be another matter. It is a lot more fun to listen to a howling pile-up with the quarry already in the log than to be wrenching out your guts calling frantically while slowly losing propagation. The intent of this book is not to help you gain an SWL DXCC, but rather one with your call on it!

None the less, pile-ups, large or small, are a fact of life in DX'ing, and often the DX'er has no choice but to enter them if he wishes to work the quarry for a needed new one. Fortunately for the skilled DX'er, most operators in pile-ups have little idea of what they are doing. Learning to be successful in the pile-up will be extensively covered, and treated as an important aspect of this book.

The fast reader can easily get through this book in an evening or so. Please, DON'T! The material in this book has been learned, mostly the hard way, over many years. There are many different ideas and techniques presented. They will be rather ineffective at best unless they are understood by the user. This really is not a book so much as it is a course, a course in becoming a successful DX'er. Read it a chapter or two, at most, at a time. It will be more meaningful to you.

Good DX'ing!

THE COMPLETE DX'ER

Table of Contents

Chapter

1 A Night on the Bands	7
2 Basic Listening	17
3 Basic Equipment Needs for DX'ing	27
4 Basic Operating Techniques	32
5 Basic Pile-Up Technique	38
6 Intermediate Listening	50
7 Intermediate Pile-up Technique	59
8 Advanced Listening	69
9 Advanced Pile-Up Technique	76
10 CQ DX CQ DX CQ DX DE...	85
11 CONTEST!	92
12 Morning Call	99
13 Summary, Section 1	108
14 QSL'ing	111
15 Night Call	119
16 Graduate Hunting	131
17 Antenna and Tower Notes	135
18 Hunting Continued	140
19 The DX'er and his Environment	145
20 Hunting - Again	153
21 Accessories for the DX'er	161
22 Special Language Techniques	168
23 Winning, Losing, and Playing the Game	179
24 DX'pedition!	182
25 The Last Secret	198
26 Conclusion	201
Appendix	203

1
A Night on the Bands

Dinner is over, and the kids are doing their homework. The XYL is at a school meeting. I don't have any homework from the office. It's a fine night for DX'ing. I get a fresh cup of coffee, and head downstairs to my basement shack.

I think about how nice it would be to have a shack upstairs, with windows, but it really isn't important, especially when band conditions are good and the DX is rolling in. Conditions were pretty good this morning; I had a brief look at the bands before I headed for the office. Long path on twenty meters was good into the Middle-East, but there was nothing on I needed.

I turn the gear on, then go through my start-up check list. Let's see here. Antenna at 45 degrees, Europe, Middle-East short path, OK. Two meter rig on the local DX net frequency. Linear tuned up for twenty meter CW. Check.

I glance at the calendar. It's May 13. But my digital GMT clock shows it's 0037, which means the GMT date is May 14. I note the GMT date in the log. I'm always careful to write the

GMT date into the log, having learned my lesson when I almost didn't get a T32 QSL I desperately needed. Only a friendly QSL manager willing to search through the logs saved me, when he found me in the log a day later than my QSL showed.

Everything's running OK. I settle the headphones over my ears, set the transceiver dial at 14,000 kHz. and begin tuning up the band.

The first signal I hear is a CQ, with a trace of flutter on it. I listen; it's a UK9 calling CQ. Nice S-6 signal, which isn't bad considering my antenna is on Europe. But I don't need Siberia, so I keep tuning on. There's a high speed CW signal, rather strong, sending in German. Pretty fist, too. Ah, he's turning it over to the other station; OK, it's DL1BU QSO'ing a YV1. And there's the YV1, not as strong, but good copy, and replying in German as well. It must be pretty late in Germany; yes, about 2:30 AM local time. I move up the band some more.

Next, I come across a VE3 calling CQ, and just above him a W5 calling VU2BK. Let's see if the VU answers him, and how strong he is. I turn the antenna straight north in anticipation. The W5 finishes his call. The VU comes right back, nice, strong, steady S-7 signal. The band IS good tonight. Wonder how fifteen is? But I'm not going to even peek till I see just what's happening on twenty first.. I keep turning the knob.

Wait, what's happening here? At least six or eight stations are calling somebody. I wonder who? They are all signing their calls only, going pretty fast, too. OK, someone's coming back. Not too strong, but good copy. Yes, that must be who they're after; yes, now he's giving a QSL manager, a G3 station. He's signing. OK, it's ZC4BI, on Cyprus. Not a bad catch, but again one I don't need. I pick up the microphone of my two meter rig, "ZC4BI, Zebra Charlie Four Bravo India, fourteen-oh-nineteen, fourteen-oh-nineteen, from W9KNI." I get no response or request for a repeat, so I continue tuning.

It must be near morning now in the Middle-East; perhaps something good will show up from there. Sure would be nice to tie into a 9K2, or to catch the ever elusive YI1BGD calling CQ from Baghdad. Guess it's a little early yet, but those fellows often do operate in the cool hours of the early morning, and the band certainly has the propagation to carry the path if one of them does show up.

I hear a raucous CQ, T7 with a chirp. OK, it's an LZ. A lot of those boys just can't get the parts necessary for a clean signal; you have to give them credit for being on the air at all, considering the circumstances they have to cope with for parts procurement. He has another station calling CQ just above him; it's a UD6. The path into the Middle-East is in fine shape. Maybe I'm going to get lucky. I glance around at the gear, focusing my eyes again on the equipment. Hey, the antenna is still pointed straight north, from when I was listening to the VU2. I ease it back around to 45 degrees. The UD6 comes up part of an S unit. I continue turning the knob, slowly, a bit at a time.

I come across a station using the Cyrillic Morse letters that are not used in English Morse interspersed with International Morse characters. Probably it's a Russian. Yes, he turns it over; it's UK9CBB working a UW3. A night like this is pure joy to listen to, with stations from all over the world coming through. My receiver dial is a magic carpet gone mad, with never an idea of what lies on the next kilohertz, only the knowledge that whoever I will hear there will likely be far away, in a strange land. Even if I don't find a new country - and I rarely do - listening to an open band is always exciting.

There's a VE3 calling CQ. Here's a PY signing clear with a ZS6. There's a weak CQ. OK, it's a WB5. Wait a minute. He's too weak to be in W5 land. Yes, he sounds like a DX signal, for sure. Darn, he quit. But it was WB5BJL portable something. Almost had it but QRN got him. Maybe he'll come back and call CQ again; I don't hear anyone calling him. I listen for at least another minute, but no sign of him. I start tuning higher up the band again.

I find another UD6, and an HA. Then an IS∅. Years ago, they were pretty scarce; I remember my early years of DX'ing, when an IS1, as they were in those days, was rare, and it took me several years to snag one. Hey, there's that WB5 again, a little higher up the band, calling CQ. OK, this time let's get his whole call. He's in the clear. Yes, it's WB5BJL portable something 7 something. Hey, it's WB5BJL portable A71! Qatar! A new one! If I get him, that is.

I punch the spot button on the transceiver, and quickly zero his frequency, then punch the synch button to lock both my transmit and receive frequencies together. (On many transceivers this is marked "A=B.") I hit the linear power switch. The surge control relay snaps home as the filaments of the 3-500Zs come to life, and the whirr of the blower becomes faintly audible. The plate voltage needle settles at 3500 volts; I'm ready. He's still calling CQ. Hope no one else has spotted him. He's pretty weak. Hope he can hear me. I carefully recheck to be sure I'm dead on his frequency. Yes, I'm ready.

There, he signs. I start calling him, with one hand on the transceiver output level. I don't want to overdrive the final. Oops, too fast on the keyer. I drop the speed a bit, trying to match the speed he used in calling CQ, as I make my call, "WB5BJL DE W9KNI W9KNI AR." I listen, holding my breath. Darn, he came back to someone else, an HB9. OK, let's copy it...

"..579 579 HR QTH DOHA DOHA QATAR ES NAME RALPH RALPH BT PSE QSL VIA CB QTH HW CPY? HB9AMO DE WB5BJL/A71 KN". OK, if he's for real it definitely is a new one for me. I listen carefully for the HB9 and find him about 200 hertz higher than the WB5. I hope the HB9 isn't waking the whole of Europe with his catch. But no, he's not making any big deal of the /A71 part of the call sign, any more than I did.

The HB9 doesn't seem interested in a ragchew with the WB5, thank goodness. The quicker I can get this fellow the better, before a real pile-up forms. The HB9 passes his QTH and name, and only a 559 signal report. Good, the A71 isn't strong in Europe. That should improve my chances; it means there's less chance that a horde of Europeans will find him. I decide not to try a tail end call; it may upset the WB5, and at this point I don't know that any one else is waiting in the wings. The HB9 is a few hundred hertz off the A71's frequency; I press the synch button again, switch to the "B" VFO, zero the HB9, then throw the VFO selector for "Receive A - Transmit B". As I do so, I am hoping a mob of hungry DX'ers is not doing the same thing. There, the HB9 is saying 73 and turning it back to the A71.

Yes, the WB5 is sending his 73, and he signs clear. They exchange dit-dits, and I call, "WB5BJL WB5BJL DE W9KNI W9KNI AR". There is certainly no sense in emphasizing the A71 part of his call; the WB5 knows what the rest of his call is, and if I don't snag him on this call, I don't want to be attracting competition. I'm not just your ordinary dummy.

I finish my call and listen, again holding my breath. Darn,

he's not coming back to anyone, but there was no one else calling when I finished my call. Wait, he's transmitting; "W9KN? W9KN? PSE AGN DE WB5BJL/A71 KN". Hot Dog! I start calling again, "WB5BJL DE W9KNI W9KNI KNI KNI KNI W9KNI AR". I hold my breath again.

Hah! All right! "W9KNI W9KNI DE WB5BJL/A71 R GM OM ES TNX CL BT UR RST 569 569 HR IN DOHA DOHA QATAR ES NAME IS RALPH RALPH BT TNX FIRST W QSO BT QRV YESTERDAY BT RUNNING 150 WATTS AND LW FROM HOTEL ROOM BT QSL CB QTH OK BT HW CPY? W9KNI DE WB5BJL/A71 KN".

"R WB5BJL DE W9KNI R FB RALPH ES TNX QSO ES FIRST A71 BT UR RST 559 559 HR NR CHICAGO NR CHICAGO ES NAME BOB BOB BT RIG HR KW ES 4 EL YAGI BT QSL OK VIA CB BT WX HR 66 F TODAY ES CLR BT HW CPY? WB5BJL DE W9KNI KN". As soon as I release the Bencher paddle, I grab the two meter mike.

"Hey, I've got a good one. WB5BJL portable Alpha 71, on 14028, 14028, I'm in QSO with him; here's W9KNI". That's sure to draw attention; there are a number of fellows in the club who need an A71, but I do not listen; I'm back to copying the WB5. "R OK BOB ES TNX INFO BT WX HR HOT HI HI GOING TO BE 120 DEGREES TDY OK BOB TNX QSO QSL SURE 73 ES KP COOL W9KNI DE WB5BJL/A71 SK".

"R WB5BJL/A71 DE W9KNI R FB RALPH ES TNX INFO ES NEW COUNTRY BT VRY EXCITED BT QSL SURE ES BEST DX 73 DE W9KNI SK". Whew! A new one! HOTTTTTT DOG! I listen on the frequency. There are four or five stations calling him, all but one sounding like locals. Guess my two meter call did attract attention. Good, that's what the two meter frequency is for, to help each other out. There, the WB5 is coming back to someone; let's see who it is. OK, it's W9DWQ, Ed. Glad he got him OK. This calls for a celebration; I decide to get another cup of coffee.

By the time I get back downstairs, the WB5 is in QSO with another of the locals, AB9E, and already they are signing clear. Guess the WB5 is speeding things up. I listen a moment; it's no wonder. There are perhaps twenty stations calling him now, and only a couple of them are local. Guess the word gets around fast, even though I'm sure none of our gang overly emphasized the /A71 aspect of the WB5's call. Of course, the careful listener can often tell if another DX'er is into a rare one; the spacing of the fellow's sending tends to get choppy, and somehow the excitement manages to work its way through the best of paddles and keyers.

Well, with that mob on frequency, and the A71 already in the log, there's nothing left there for me, but conditions are still good. I better keep moving. Maybe I have the hot hand tonight. I start turning the knob again. Onward, ever onward.

Hmm. There's a UL7 with a 589 signal. What time is it? OK, it's 0217 Zulu. Gosh, it was pretty early in the morning for that A71. I'm still grinning over that one; a rare country gracing the log book, an all time new one for me. Sure am glad I dug that WB5 out of the mud. I'll get the QSL off tomorrow. What's that weak CQ? OK, a YU3. There's another CQ above him. Ah, it's 3B8DA. Guess the sun must be rising in the Indian Ocean, OK. Nice signal, too. But another one I don't need. Keep turning the dial.

I finish going up through the band, move back down to 14,000, and begin tuning higher again. What's that? "5NN K". Clean, sharp fist, about 30 words per minute. There it is again, "R SK UP 5 K". This sounds promising. Let's see who it is. There's no one calling him on his frequency. No, there's a station signing his call. And a traffic cop telling him to QSY. And there went the DX station, but completely buried by the cop. Oh, if only those traffic cops would hand out their citations a half kilohertz off.

Ah, the cop has quit. And there goes the unknown DX again, "N4AR 5NN K". I punch the synch button, locking my two VFO's together, then pop the "A" VFO five kHz. and look for N4AR. I don't spot him within a few seconds, so I throw the VFO switch back to the "B" VFO. Yes, there's the DX station. "R SK QRZ UP 5". I flip the VFO switch again, looking again five kilohertz up. Wow, what a mob scene. It seems like there must be hundreds of stations calling, but who are they chasing? I flip the VFO switch again.

Yes, he's transmitting again. "K8CW 5NN K". I jump back up again. There are still plenty calling him, and within the few seconds allotted, I fail to find K8CW. I switch back down. Yes, there he is again. "R 73 QSL DJ6SI QRZ DE T55SI UP 5 K". OK, great. But what the heck is a T55? I grab my "Second Op", and take a look at the ITU call block allocation. Lemesee here. Yes, there it is. T5A-T5Z. Somali Democratic Republic.

Hey! That's what used to be a 6O1. And I need that one too!

He comes back to another station. I miss the call of who he is working, so I wait, with my transmit frequency now set five up, as instructed. There, "R SK UP 5 K". I start to call. No, that's too slow; the keyer is still fairly slow for the WB5/A71. I quickly crank the speed control up a bit. I finish my call and listen again. "1HZ 5NN K". I missed the prefix of who ever he's working now. This fellow is a fast worker. I jump the receiver back up five. No luck. I tune a little higher, but still nothing. Then I jump back to his transmit frequency; at the speed he's logging QSO's I can't afford to hang around waiting.

"UP 5 K". Yes, he was just finishing. I call again; "DE W9KNI K". and listen again. Yes, there he goes; "W2BA 5NN K". I know 2BA; he usually has a good signal, and he uses a bug instead of a keyer, so his fist is a bit more distinctive. I flip the VFO switch to look for BA's transmit frequency. Nothing at five up. I start checking higher in the band. I take a very brief look up to around ten kHz. higher, but without any luck. I flip the transceiver VFO back to the T55's frequency.

Yup, there he goes, "R 73 JACK SK QRZ DE T55SI QRZ UP 5 K". I move my transmit frequency ten kHz above him, and call blindly, "DE W9KNI K". I don't give him my call two times; clearly he is going with people who sign their call only once; which is one of the characteristics of a really good DX'pedition operator. I listen again. "W4QQN 5NN K". I pop back up again, looking for W4QQN, checking around ten kilohertz higher. Nothing.

This is getting ridiculous, not to mention frustrating. I know I could have heard W2BA if I could have found him, and the same is true of W4QQN; he's no slouch with a signal either. Clearly the DJ6 is listening other than the stated five kHz. up. In fact, I decide, he's probably listening up rather more than five, since I'm pretty sure he's not listening less than five. The "five" must be an indication that the minimum is five, and especially it means, "Don't call me on my frequency." Wish some of the traffic cops would take notice of that as well.

I skip back to his frequency. Yes, there he goes, "R SK QRZ UP 5 K". I bump my transmit frequency to a spot 14 kHz above him, on a hunch, and give him a fast, "DE W9KNI K". I listen. He comes back; "WB4OSN 5NN K". I switch back to my transmit frequency and start quickly tuning higher. Hah!

THE COMPLETE DX'ER 15

There! "5NN 73 DE WB4OSN SK"! I switch VFO's, leaving my transmit frequency on WB4OSN's frequency. The T55 signs clear, and I call, then listen on the T55's frequency. There, he comes back, "W4QM 5NN DALE K".

I switch back to my transmit frequency. Nothing. I tune a bit higher. Yes, there, about one higher, "4QM SK," about twenty kilohertz above the T55's frequency. I switch back to the T55, and call when he clears. I finish my call again, and listen. "K4KQ 5NN K" I switch again to listen on my transmit frequency. Yes, there's K4KQ, about three quarters of a kHz. above my transmit frequency. Hmm. It looks like the T55 is moving up after every contact. I can play that game; I move up three quarters of a kilohertz above 4KQ. I switch again to the T55's frequency. Yes, he's transmitting, "UP 5 K".

Again I call. I listen; he comes back. "K5LM 5NN K". I go back up with my receiver to my transmit frequency. Yes, there's K5LM, virtually dead zero on my transmit frequency. I set up another three quarters of a kHz. higher. He clears. I call. He comes back again to W9VNE. I listen. Again, I'm almost dead zero on VNE'S frequency. I start to feel confident; I'm almost sure to get him within the next few calls, now that I've figured out his listening pattern. I hop back to the T55's transmit frequency.

"R SK QST QST MUST QRT NW FOR BED BT WILL BE QRV AGN TMW 73 DE T55SI SK CL".

ARRRRRGH!! Curses, foiled again. The story of my life, a minute late and a call short. But I quickly cool down. Who am I to complain? I've already gotten a new one tonight, a nice rare one, and besides, the T55 said he would be back on again tomorrow. And, I've got his pattern figured out pretty well; hopefully it shouldn't take long to nail him when he does come on tomorrow. He certainly is a fine operator. And Hope Springs Eternal.

I start tuning up the band again. It's pretty cluttered above the T55's frequency; there are still a lot of stations calling him that didn't get the message. But soon I'm clear of them, and the band is quieter. I glance at my clock. Huh, it's already nearly 0300 Zulu. The sun is definitely up now in the Middle East, and must be so as well in Somalia, where the T55 is. Guess he must have been up all night, and intends to sleep during the daylight hours.

I hear my wife's voice upstairs; guess she must be back

from the school meeting. I decide to go upstairs to chat with her a few minutes, and to brag about my new one in Qatar. I'll have to get the atlas out and show her where it is...

...In about 20 minutes I'm back in the shack. I decide to start tuning again from the bottom of the band.

The band seems quieter now. I hear no Europeans at all. Though most of them are certainly in bed, I would expect to hear one or two if we had propagation, so we must have lost the path while I was upstairs.

I slowly work my way up the band, but it does go faster with fewer signals to check out. I keep hoping for a good African or someone in the Middle-East, but I seek in vain. Guess we really have lost the path. No, there's a 9J2 working a G, but that path is considerably easier to hold than one to the further north areas. I listen, but when the 9J2 turns it back I can't hear a trace of the G station. But I can't complain; I got what I wanted.

I swing the antenna further to the south, for a quick look for FT5WK on Crozet, who the bulletins have listed once or twice in the last few months. I don't have much hope, though; the fellow has shown only rarely, and in a very inconsistent pattern, the kind that makes a DXer's nightmare. I decide to move the antenna to bisect the paths to the VP8/LU-Z islands and the paths to the FT islands. A move across the band yields nothing, however. It's getting late, now, 0348 Zulu, 10:48 PM CDST, and I have to rise early tomorrow for a breakfast meeting. Time to pull the switch.

I leave the shack without complaint. The A71 was a fine new one I did not expect, and almost a sure QSL. Too bad I didn't nail the T55, but I wasn't expecting him either, and he'll be back tomorrow. And I already have a handle on his operating pattern, which is sure to help. Even if propagation is off, that should be a decent path, and I should be able to manage it OK. At the top of the stairs, I switch off the basement lights, feeling rewarded. Besides, it was fun.

2
Basic Listening

If your primary interest in DX'ing lies in reaching the DXCC Honor Roll, the first and most important skill you must have or must learn is how to listen. No other skill is as important for the DX'er.

Making a good start at learning how to listen is easy, fortunately, and indeed, it is fun. You are even almost guaranteed to work new ones in the process. And, you never will really stop learning how to listen; your skills get sharper from experience as time goes on. You may be puzzled at this point about listening. After all, isn't it a matter of just tuning the transceiver?

Of course, to a certain extent it is. However, the art of listening is also learning to let nothing get by you as you tune a band; a 6th sense about what band to be listening on, and knowing what heading to turn your antenna to if you have a rotary array. Listening is finding out about schedules with rare ones being traded by other stations who would just as soon you weren't listening. Listening is watching the inter-Pacific traffic nets at 3AM so you can nail the elusive KC6 after the net is over. Listening is knowing when to listen, and when to not waste your time, especially when you have other obligations that also must be met.

Again, listening is the mental work of digging calls out of fierce QRM, or reading extremely weak signals. These, too, are skills that improve considerably with simple practice. It is the joy of taking the pulse of an open band, and hearing the world.

But it takes practice, and an understanding of what you, as a DX'er, are trying to accomplish. Two amateurs can sit down, side by side, at two identical receivers, using identical antennas listening to the same open band. The experienced and skilled listener will unearth country after country as he tunes across the band, digging rare ones out, exploring, probing; while the casual listener sitting at his side will, spending the same amount of time and with the same objective, hear little or nothing of what's really going on.

But listening skills can be learned; indeed, as mentioned above, a great deal can be learned in a rather short time. In fact, if you have not already acquired some of the skills, once you have read and understood this chapter, you will almost instantly be twice as effective as a listener.

Before we turn the transceiver on at all, let's discuss a little of the hardware, and a few ground rules. First, let's check out your headphones. Good listening cannot be done effectively with a speaker; you must use headphones. Since, presumably, you are going to be wearing headphones a good deal, you want to insure that they fit your head comfortably, and display no tendency to fall off. On the other hand, they should not clamp a death-grip on your head either. They should cut down outside noise considerably, but not so much so that you can't hear the family call you, the phone ring, or the rare one called in on the two meter DX net. Also, they should let your ears breathe a bit; sweaty ears are not very comfortable, though at times that condition may be unavoidable.

There are several brands of communications headsets offered to the amateur, of which several models are excellent. Curiously, the most expensive are not necessarily the best. And some of the very best are not readily available to the amateur at all. You might check out what the local flight service shop at the airport has to offer. Remember, you want them to be comfortable, and to sound good. Availability of replacement ear pads is a nice plus.

There are many excellent, modestly priced stereo headphones available today, and, since they are designed for hours of listening to music, they tend to be fairly comfortable. Stereo headphones do suffer one disadvantage, in that they are designed for high fidelity listening, usually somewhere between 50 and 20,000 hertz. Amateur communication bandwidths for CW usually run from about 250 hertz to 1000 hertz, while SSB communications run typically from 300 to 3000 hertz. Unfortunately, most amateur transceivers generate white noise components that often go past 10,000 hertz, and this noise can be very fatiguing for the DX'ing listener, especially over long operating sessions.

If you find a pair of stereo headphones you fancy, there are several ways to eliminate the white noise. A simple toroidal passive audio low-pass filter can be built. Many DX'ers use active audio filters which reduce such noise, and offer other benefits as well. At the least, a by-pass capacitor can be placed across the audio output line for some roll-off of the high-frequency white noise component. The value for such a capacitor must be derived experimentally, and is, of course, significantly affected by the impedances involved. However, with low impedance 'phones, a value around 10 microfarads makes a good starting point for experimentation.

Since the hearing response of each of us is likely to be different from that of the next ham, headphones really should be tried out plugged into YOUR rig before finalizing a purchase. Check them out for comfort, fit, and for the pleasantness of the output into your ear. Most dealers are willing to allow a return on such items if the purchased headset is not comfortable for you. But ask about a possible return before the sale is final! And if granted, don't abuse the privilege. Make life nice for the next ham as well.

Next, take a look at the operator's chair. Since a good deal of time is likely to be spent in it, it should be a comfortable one. Your author uses a hardwood captain's chair, purchased cheap because it did not belong to a matched dining room set. It has proven very comfortable over the years. Many people choose soft chairs, but these may prove to be less comfortable over an extended operating session. In any case, care should be taken to insure that a chair is not going to stand in the way of operating pleasure.

Now, before you actually start tuning, there is one rule for your own good you should follow. Always do general tuning with the AVC or AGC ON! Old time CW operators will tell you that the way to tune is with the AVC off, and one hand riding the RF gain control. They fail to tell you about Tinnitus, also known as ringing in the ears, which is a malady sure to shorten a DXer's career should he come down with it. Tinnitus is caused by excessive exposure to high levels of static or very loud CW signals. A proper AVC circuit will protect you from most of this, along with an awareness of what could happen and a dose of common sense. Like, don't listen to 80 meters in mid-summer with the gain wide open.

The basic premise of learning how to listen is simple - you must identify every signal you hear. This does not mean you have to necessarily know the call letters of each station heard, but it does mean that you have to know whether he would be a new country for you. For example, if he's S-9, and saying that his QTH is Mansfield, Ohio, you might want to tune on. But, even then, you want to be sure that the station that he is in QSO with is not A51PN. Unless you have the A51's card on the wall. But, as we will discuss more later, if he is in QSO with the A51, his fist or his voice will usually show it, and we would detect clues that suggest we should hang around and see what gives.

The ideal situation for a hungry DX'er is to find a station from a needed country calling CQ that no one else happens to hear. Then a simple one by two call is usually all it takes to put the DX into the log, even though the DX'er may not have one of the more outstanding signals on the band. Consider three facts. The first is that the DX station must do something to initiate a contact when he gets on the air. Unless he is keeping a schedule, he has two choices - either he calls CQ, or he calls another station who is calling CQ or who is finishing a QSO. Mr. Rare DX does NOT start his day on the band with an automatic pile-up. Pile-ups grow as more people are drawn to a frequency by the pile-up, rather than by the DX operator's signal itself.

The trick is to catch the rare DX sending his first CQ, or at least finishing his first QSO, rather than tripping over him as he is pursued by a screaming pile-up offering the DX'er little chance of success.

The second fact is that if some one else does hear the DX station calling CQ at the same time you do, at least the pile-up you are in is composed of two people - you and the other station. All else being equal, you have a fifty-fifty chance against the other operator. Contrast that with your chances in a pile-up of a hundred stations, where, if everything is equal you have one chance in a hundred.

Of course, all things are NOT equal, but hopefully the point is clear. If you find a station in a new country by dint of careful listening, you have a far better chance of working him than if you find him under siege from a howling pile-up.

The third fact worthy of note is that the signal strength of most rare DX stations is average at best. The rare DX station just isn't likely to go to much trouble to develop a crushing signal into the States - he assuredly already has "W" confirmed unless he's very new. So, to find him, especially when he's calling that day's first CQ, it's up to you to do the digging to scrape him out of the background.

Most DX'ers searching for new countries do not tune carefully, RF gain up, listening to every signal. Hence, they fail to hear that first CQ. They fail to find him in QSO with another station. They don't find him at all until a pile-up forms on the DX station, showing them the way. Tuning for the pile-ups is a great way to rack up an SWL DXCC, but it is a sure prescription for an acid stomach as well, especially if you do not have one of the dominating signals on the band. Patience is indeed a virtue, and one that the good DX'er needs plenty of in any case, but having the new one already in the log when the pile-up forms beats the heck out of having to be patient and hoping you can outsmart the gang.

Remember, pile-ups don't usually start before the DX operator arrives, and few DX stations always show up at an exact time on an exact frequency. (Except for schedules.) Careful tuning will put DX stations into the log at a far faster rate than any other technique, and will give you the satisfaction of listening to howling pile-ups on stations you were the first to unearth calling CQ. It's a nice feeling...

You say that twenty is open nicely, and we have a bit of time for tuning? Let's go have a look around...

Ohhh-kay. The antenna is at 45 degrees, the direct shot at Europe and the Middle-East, and within a half S-unit of being optimized for the trans-polar Central Asians, and for the North and Central African states as well, a rich lode for the hungry prospector. We set the receiver at the bottom edge of the band, turn the RF gain up till the background noise makes a nice soft backdrop of band hiss and start slowly moving up the band.

There's our first signal; pretty loud. In QSO with someone. Clean, very steady signal; probably a W. Ragchewing; yes, he's asking the other fellow how his strawberries are doing. Somehow I rather doubt that A51PN is party to either side of this QSO. Ahh, he's turning it back. OK, he's working a VP9 and it's a W3. On we go. There's someone calling CQ, no where near as strong as the W3, but still excellent copy. Steady signal, there, he's signing his call, OK, it's F3AT. On we go. There's a signal going rather fast; at least 35 words per minute. And in German. Good strong signal, S-8 or 9. Not quite as steady, bits of rapid QSB up and down. Probably a European a bit further north would be my guess. The nearer the pole, the more unstable the signal, it seems like, unless conditions are exceptionally steady. Ah, there, he's turning it over. Right, it's YO3AC working a PY in Brazil.

There's a rather weak signal just above the YO/PY QSO. Calling a W1. A G? No, G something, I think, but I'm not sure. Copied an 8 in his call though. Darn, almost had it. Wait, was that a PM8? Probably not. Phooey, he quit calling. Let's wait a moment and see what develops. No one is coming back to him. Yes, there he goes again, calling the W1 again. Sounds like a schedule. There's his call again. Yes, got it. GW8PG. OK, Wales. Nice, but I don't need it. Onward, ever onward.

Listen to that loud signal calling CQ. Sure would be surprised if that was a DX station. OK, he's signing. Well I'll be darned! It's FP5PS! I have one, in the log and on the wall, but they aren't exactly garden variety either. Just goes to show - even loud signals from nearby can be DX. Let's wait and see what kind of a pile-up he attracts. With a signal like that, he's sure to attract a big following in a hurry. There, he's signing. This ought to be good.

Huh. There's no one coming back to him. No, wait, there's

a weak signal calling him. Giving him a one by two. A WB7. Yes, the FP5's coming back to him. Bet the WB7's pretty pleased; from the sound of his signal he can't be running much soup. Well, maybe it's a new one for the WB7, but it sure isn't for me, and sitting around here listening to the QSO and speculating doesn't do my log book any good. Time to be moving on.

Ah, another QSO going on. Sounds like a DX signal for sure. Nope, he's saying QTH HR MADISON WIS MADISON WIS... must be a W9. The backscatter on his weak signal fooled me. Yes, OK, he's working an LA. I keep going. Listen, there's someone calling CQ. Not too strong, but nice clean signal. Yes, a little flutter, generally the characteristic of a DX station. Yup, an HB9. Keep tuning. Catch that; another QRQ ragchew from the sound of it. Hmm. WX HR WET ES COOL, BEEN A WET SPRING IN LONDON... OK, nothing there for me; wonder who he's in QSO with; probably a W. He's pretty loud and he's ragchewing, a good sign that he's got a yagi or a quad, and that it is pointed to the States. Hey, I'm almost right; he turned it back to a VE3.

Ahhh. Some turkey is tuning up. Taking his sweet time about it too. Shouldn't require more than a few seconds of on the air time. Try to tell him that, though. See how his signal goes up and down in strength as he twiddles the knobs. Must be rich; new finals aren't cheap these days. Do I hear a weak signal under him? Yes, there's someone calling CQ, but I can only hear him when that turkey is out of resonance and his signal drops. There, he quits. Probably blew a fuse when the plate sagged into the screen. Served him right if it did. Dummy loads aren't that expensive. But the other station that was calling CQ is gone now too. Darn; I wonder if it was a good one?

No one is calling anyone, guess that carrier wiped out everyone. Let's hang around a few more seconds; maybe the CQ station will call again. Yup! There he goes again. Not very strong; I'm surprised I heard him at all under that carrier. What's that call sign? An OZ5? No, that isn't right. Hey! It's OY5FR! The Faroe Islands! That's one I need!

Quickly, I press the spotting button and tune the dial to match the note, then punch the synch VFO button. Check the keyer; yes, speed's about the same as his. Aghh. He's still calling CQ; the whole world will have found him by the time he's through. Ah, at last, he's signing. I pause a tiny second; whew, no one else is calling him.

I give him a one by two, and anxiously turn it back. There! The sweetest music in the world; my call coming back from his rig. H-o-t-t-t D-o-g! Quick, note the GMT time, before I forget when the QSO started. All right!

I finish quickly with the OY, sitting there with a big grin on my face. That was pretty neat, the way we spotted the OY underneath the idiot tuning up; and then waiting him out to hear his call. Of course, by all rights, it should have been a G, or a DL, instead of a nice new country. But it wasn't! It was a new one. And we got him, by careful listening. But the day is not done. Let's tune some more. Maybe we'll get lucky again.

I listen one more minute on the frequency of the OY5. Many stations are calling him now, doubtless attracted by my signal when I worked him. But he seems to be gone now, probably put off by the small pileup he spawned. How sweet it is! I start moving the dial again, marching slowly up the band.

It still sounds the same. There's a DJ calling QRZ? I listen; a KP4 answers him. A W4 opens up with a CQ DX. Let's pause a moment and see what he snags; his signal is nice and loud here. He sends a standard three by three pattern CQ, and signs. I wait with him. Yes, several stations are calling him; an HA, a YU and an IT9. I move on.

There's a weak signal, a bit chirpy, just about buried in the noise. I strain my ears, and advance the RF gain a bit higher. Still hard to copy. I crank in a bit more selectivity from the audio filter; yes that helped a bit. Move the dial a hair, to perhaps center him a bit better in the transceiver passband. Yes, that did help a bit more. What's he sending? That's not standard international morse. Oh, OK, must be Cyrillic code; international Morse with extra elements for the Cyrillic alphabet. Probably a Soviet station; only they, the Bulgarians and some of the YU boys use the Cyrillic alphabet. But that's OK, I still need several Soviet republics.

I keep listening; the station keeps sending without signing, working break-in with his contact. But soon I hear the "DSW" and I know he is about to sign. Yes, it's UK6GAB. I don't need to grab my Soviet prefix list to find it's one that I need - it's the same as a UG6, one I certainly need. Needless to say, I'm setting up for him as fast as I can go. I wait; he has signed clear, but I think that the other station he was in QSO with is still signing. Wait, there is the traditional "dit-dit" from the UG6. I call him, giving a one by two call.

I listen, waiting with bated breath. Nothing. I wait a few more seconds. Still nothing. I call again, a bit slower this time, and with a two by three call. Still nothing. I try a three by four call; again to no avail. Looks like I've lost out on that one. Well, he wasn't very strong; perhaps he didn't hear me. But I did have a clear shot at him; perhaps next time my luck will be better. I start tuning up the band again.

There's a fairly loud signal calling CQ. But he has traces of flutter on his signal, and no sign of back scatter. Betcha he's DX. Good S-9 signal. Yes, it's DL1BU. Let's sit tight a minute and see if any one comes back to him; he's ending his CQ. Hah, listen to that! There must be five or six stations calling him, all W's from the sound of it. Funny; I've heard several fairly rare DX calls on the band, while a DL is common. Yet, the DL1 is the station getting the attention, while the rarer ones get hardly any interest.

But there's one other difference. The DL is fairly strong, and with a good fist. The other stations were much weaker, except the FP5, and he was so strong he sounded like a local. Hmmm, maybe there's a lesson in that; perhaps most DX'ers on the one hand don't dig into the noise a bit, and on the other hand don't pay much attention to really strong signals either. But they sure can miss a lot that way. Anyhow, that's their problem. I'm going to keep tuning.

Hey, there's a bit of a pile-up, just above the DL. It's all Europeans from the sound of it. Not that many of them, though. OK, someone loud is coming back to one of them; yes, it's XE2MX, off the back of my antenna. Some of those fellows are still calling him. Listen to that one with the chirp, that 7X2. Hey, wait a minute, I need a 7X2; that's Algeria.

But I have a dilemma. What should I do? The 7X is after the XE; he's not likely to take kindly to my calling him and attracting attention to him when he's after the XE. The XE is rare for him, probably a new one, and probably a new zone, too. And I can be fairly sure a W9 call isn't going to send shivers down his spine. I'd better wait a bit and see what gives; maybe something will happen so I'll get a better chance of working him instead of getting him mad at me.

I listen to the XE; he obviously does not know that the 7X2 is calling him, or possibly he doesn't care; he's working an

HB9. Sure hope that the 7X doesn't give up and go away, but if I call him now I'm taking a serious chance of angering him, and that's not good. The XE turns it back to the HB; I decide to wait and see what transpires.

Soon, the XE signs clear, and the European pile-up starts again, with more stations calling furiously. I listen carefully for the 7X2, but hear his distinctive chirp only a couple times when the stronger European signals stop transmitting for a moment. Not much chance that the XE is going to hear the 7X out of that mob. I wait for the XE to come back.

I keep waiting, but the XE does not show. I listen a bit longer; the pile-up starts to taper off, as it becomes apparent that he is not coming back. As it thins out, I hear the 7X better, with less QRM on him. I set my transmit frequency a hair above the frequency of the Mexican station and wait; I hear the 7X call again. His call is longer this time; probably he's about to throw in the towel. He signs; I wait a counted out five seconds, and, with no sign of the Mexican station, I call him, giving him a brief one by two. (His call once, mine twice, as in "7X2BC DE W9KNI W9KNI AR")

I listen, I hear "KNI QRX QRX XE2MX DE 7X2BC 7X2BC 7X2BC AR" Hey hey! Maybe I've got him. At least he heard me. I listen, anxiously. There's someone calling the 7X, but it doesn't sound like the XE; no, it's a WA4 something. Gosh, hope I'm OK and that the WA4 doesn't upset my apple cart by getting the 7X2 mad. The WA4 signs. I wait, again holding my breath.

Yes! "W9KNI DE 7X2BC R..." Another new one! I start putting the happy details into the log book...

And so it goes. The important lesson here is, again, to listen, trying to let nothing get past you. Listen to every signal, strong and weak, and try to identify the country it comes from, and who the QSO is with. Dig deep for the weak signals; even though you don't have the best signal on the band, experience will prove to you that many DX stations hear better than their transmitted signal would suggest. Remember too, that even if your signal going back to them is weak, their recognition factor of their own call coming back to them is very strong, and they will try to drag your signal out of the mud. Especially if you are the only one calling.

We will further discuss listening in a later chapter, but you should be off to a good start now. Good hunting!

3
Basic Equipment Needs for DX'ing

Amateur radio today is certainly an equipment oriented hobby, much like photography, micro-computers or audio. What amateur wouldn't like to own the latest new transceiver with all the bells and whistles? If you can afford one, and find one you like, by all means buy it. But the purpose of this chapter is two-fold: to help the budding DX'er identify the type of equipment he should have, and more importantly to get across the idea that one need NOT own the finest and most expensive equipment available to be a successful DX'er.

Permit me to demonstrate this further by comparison to a very common situation in amateur photography. Jack really enjoys photography, and wants to take great pictures like those of George, whose skill he admires. He sees that George takes his great pictures using a Super-Whizz-Bang Gargleflex with a 37.5 to 98.6mm turbo-zoom Coketar lens, and 43 filters. So,

Jack saves up and buys, at considerable expense, a Super-Whiz-Bang Gargleflex, the referenced Coketar lens and the 43 filters. (Jack has always admired Sam's pictures as well, but considers Sam's success a matter of luck or some similar aberration, since Sam uses a very old non-automatic camera.)

But, if anything, Jack's pictures with the Gargleflex are worse than they were before using his old camera. Jack is very disappointed, but soon notices that Pete takes really superb pictures using a Kablamoblitz 80 with a selection of fixed focal length Accublah lenses. Sure enough, Jack is last seen getting quotes on the trade in of his Super-Whizz-Bang for a Kablamoblitz. And, his pictures will probably be worse yet.

Some photographers are natural successes at it; others have to learn. And it can be taught. For a whole lot less than the price of a Super Kablamoflex. I'm sure the parallel is obvious; don't go refit the shack just because you are not working DX. It won't help. Instead, figure out what you need to change in your operating habits so that you will become an effective DX'er. A DX'er can and will work DX with two watts to a random wire. So can you.

Rather than getting involved with specific pieces of equipment and naming names, generally speaking, we will explore types of equipment, desirable features, and a general philosophy of how equipment relates to the DX'er and his needs. Let us presume, for the purposes of discussion, that you have no equipment, and propose to outfit a station from scratch.

The first question, once we have decided how much cash is available, is how to best spend it for maximum performance. In days gone by, the DX'er had to choose between a transceiver, generally with an external VFO, and the combination of a transmitter and receiver. A good receiver and transmitter combination was a very effective DX'ing tool, especially for the CW operator. But times change, and today it is virtually impossible to buy a separate transmitter/receiver combination.

There are a number of good transceivers available today, most of which offer the full range of facilities and capabilities the DX'er will require. The DX'er should insist that a transceiver being considered for purchase includes such niceties as built in provision for at least one and preferably two narrow CW filters, one or another method for accurate CW netting (or zero-beating,) two or more VFO's and a speech processor. However, most transceivers marketed today offers such features. Indeed, almost any transceiver which includes these features is capable of carrying a new DX'er to the DXCC Honor Roll.

The one consideration to be wary of, unless you are dedicated to SSB DX'ing exclusively, is to see that the transceiver being offered allows accurate and intuitive zero-beating (or netting) of a station on CW. To check this point out, have the netting capability of the rig demonstrated, or read the instruction manual and try out its recommendation on the method of zero-ing a station or a frequency. If no mention is made of zero-beating or netting in the manual, or if the procedure suggested is contorted and non-intuitive, that transceiver is probably a bad choice.

It is possible to accurately zero-beat a CW signal with almost any transceiver, but the techniques required for some units are so crude or so complex that no self respecting CW DX'er would ever consider their purchase. Fortunately, most of the more recently introduced models feature improved flexibility for proper netting techniques. And, should you own one of the earlier units which are more difficult to use, some special techniques will be discussed in the chapter on CW which allow easier netting than the design envisioned.

For the DX'er who is primarily or exclusively an SSB operator, the most important considerations are that the transceiver offers at least two VFO's, and features some form of RF speech processing. Even a few years ago, an RF processor was considered an exotic high tech DX tool; today virtually every rig sold features both ALC and some form of RF clipping or compression built right into the standard circuitry.

Another consideration vital to the DX'er is dependability; you can't work a major DX'pedition with your rig out being repaired! Before the purchase of any new rig, the amateur should carefully check out the reputation of the proposed rig, plus what service arrangements might be available.

DXer's of modest means should not be put off by all the talk of shiny new rigs; there are marvelous bargains to be had in used equipment these days. Most receivers, transmitters and transceivers built in the last ten to fifteen years, if in proper operating condition, offer specifications of such things as rf sensitivity, selectivity, stability and power output that compare favorably with gear of similar quality built today. Such gear is usually offered at very low cost compared to the latest gear, yet can be very effective for DX'ing. Do a little checking before making such a purchase, however. Some rigs offered, even by current well-known makers, tended to be dogs with poor reputations.

Used equipment can offer an attractive way to gain an improved operating capability. Before buying a used rig, the wise amateur will insure that one way or another service is available. Many amateurs are able to repair electronic equipment, but current design poses a dilemma. The more recently built transceivers on the market make so much use of special integrated circuits that repair, even with sophisticated test equipment, is virtually impossible without the close order support of the manufacturer, an inventory of special parts, fixtures and manuals, and the time to get parts ordered and received. This does not mean that the amateur should not purchase the late model used rig, but rather that he should first confirm that service for the rig is available, and that the manufacturer of the rig is still in business.

Older designs, while lacking some of the more modern features, can much more easily be serviced by an owner possessing technical skills, generally with only simple test equipment. As a result, many of the older rigs still find favor at their owner's operating position. However, if you lack the competence to repair such equipment yourself, you may have a worse time keeping it going than you would with a newer rig.

Most DX'ers, if they can afford it, try to keep back up equipment available. Often, the back up is the older piece of gear that has been replaced by a newer model.

The biggest question most fledging DXer's must meet as they are assembling a station is how additional funds should be allocated after purchase of the basic equipment. Of course, a number of accessories will be required, such as headphones, a CW filter, a keyer and paddle etc, and these items are treated in other sections of this book. Once they are in hand, the classic argument is whether a linear amplifier or a directional gain antenna should be next on the list.

Conventional wisdom favors the improved antenna, and in most cases that is certainly the correct decision; the gain antenna will help your signal nearly as much as a linear will, with reduced chance of RFI, and will help greatly in hearing the DX stations on the bands for which the antenna is intended.

If, however, the prime goal of the DX'er is, for example, the attainment of the Five Band DXCC award, he would likely be better off buying the linear. It will be a lot easier to work one hundred countries on the higher bands using simple antennas and full legal power than it will be to get the hundred on eighty meters barefoot, with an idle twenty meter array outside!

Antennas will be considered as a separate subject later in this book, as will accessories useful to the DX'er. If you are on the edge of purchasing something, skip ahead and take a look; otherwise we'll get there in due course!

4
Basic Operating Techniques

There are several points, techniques and procedures referenced in the preceding text that the newer operator will likely not understand fully. This chapter seeks to clear up a few of these points. Equally, more experienced operators might just want to take a peek through here to insure that we are on the same wavelength.

The first point is the term widely used by DX'ers called "zero-beating," or "netting," one of the most important tools of the effective CW DX'er. Most simply put, zero-beating means to place your transmit frequency precisely on the same frequency as that of another station, the one you are zero-beating. It does NOT necessarily mean that you tune your receiver so that the audio note of a particular station goes to zero hertz, although that is an acceptable, if crude, way of zero-beating another station. It does mean, generally, matching a spotting tone generated by your transceiver, or the local reception of the VFO of a your transmitter, with the tone of the received station.

Before even trying to zero another station, one prior step or precaution is vital. The receiver - or transceiver - MUST be set for single signal reception. Single signal reception is a term used with and for all receivers or transceivers except those with direct conversion reception. (Virtually all transceivers and receivers feature superhet design, rather than direct conversion.) The idea is that you have the receiver passband set so that it is on either the upper or lower side of the injection BFO, and with adequate selectivity to eliminate the other side. In other words, when you tune through a CW signal, as the tone of the received station goes lower towards zero, you must not hear the signal coming back up again from zero as you tune through the zero center frequency of the CW signal and "up" the other side.

Some equipment gives you no option; when you are in the CW receive mode, the passband tuning is set in such a way you cannot err. But many transceivers feature passband tuning, allowing the bandwidth of the IF crystal filter to be swept across the center BFO frequency. And you soon learn that the passband tuning must be set on one side of zero to copy upper sideband SSB, and on the other side for lower sideband SSB reception.

Passband tuning is a valuable feature in a transceiver, making crystal filters far more effective when the going gets tough, but improperly used, passband tuning can get the operator in trouble in terms of zero-beating.

If you are using a separate transmitter/receiver combination, whichever passband you choose to listen to is a matter of personal preference, although most DX'ers choose the upper side of zero beat - the same side as upper sideband - to give a little extra margin of tolerance when working near the bottom edge of the band. But, if you are using a transceiver, it is imperative that you learn which side of the pass band the transmitter frequency will be set for, and always tune CW with the passband set on that side of center. For reasons that have never been apparent to the author, virtually all recently designed transceivers transmit CW on the lower side. In any case, check the transceiver manual, and be sure your passband setting is on the correct side. If you don't, even though you use the following techniques flawlessly, you will always be transmitting a minimum of at least a half kilohertz off in real terms, enough to guarantee failure when calling other stations.

Now that we have that point established, let's get on with the business of actually zero beating a station. Let's take the simplest case first - zeroing with a combination transmitter/receiver...

OK, you have just found 5R8AL calling CQ. Now, move the transmitter VFO to the approximate frequency of the 5R8. Hit the spotting switch on the transmitter. You should hear the tone of the spotting signal. If not, jiggle the VFO dial until you do. Now, simply move your VFO dial until the spotting signal exactly duplicates the tone of the 5R8. In other words, the VFO signal of the transmitter covers, or drowns out the the 5R8. Let go of the spotting switch, and there he is again. You have set your transmit frequency dead on him, the best place to answer his CQ.

Different transceivers offer different methods of zero-beating, and unfortunately, a few offer nothing at all. Transceivers that do offer a definite facility almost always feature some sort of button that generates in the receiver a tone equal to the offset of the transmit frequency. Where such is the case, it is only necessary to move the receiver so that the introduced spotting tone is identical to the tone of the desired station you wish to zero.

Other transceivers have no formal spotting button, but are designed so that the audio side tone of the CW monitor within the transceiver is set to the same frequency as the CW offset differential of the transmitter. In such cases, generally the MOX or VOX transmit switches are temporarily switched off, the keyer is touched to generate a series of dots in the sidetone monitor, and the receiver is moved until the received station is identical in tone to the monitor tone. The transceiver is now zeroed with the DX, and the VOX or MOX switch is again turned on.

Should your transceiver have an RIT control, note that it is very important that you vary the transceiver frequency for this type of zero-beating by tuning the main dial, NOT the RIT control. RIT MUST be disabled when you are zeroing a station if you are using a transceiver. After the station is zeroed, the RIT may again be used, but remember to always disable it whenever you are zeroing or netting a station.

If you have a transceiver using digital frequency generation, it likely has an "A=B" or "Synch" button. Once you have zeroed the desired station, press that button, locking the transmitter onto the received station. You can then set the frequency mode so that one VFO is set for receive and the other set for transmit, and tune the receiver without fear of moving the transmitter frequency.

If you have any question about zero beating, be sure to resolve it as quickly as possible, and before going farther in this text. Going after DX on CW without the proven ability to zero-beat is sure to lead to considerable frustration and lost opportunities. If possible, get a couple friends to help you on the air. Have one transmit, and attempt to zero beat his signal. Have the second friend report on how close you are to the first station's frequency. Or, use a friend who has a separate transmitter - receiver combination. Simply get into QSO with him on the air, and zero-beat him using the above referenced techniques. Then ask him how close you are. He will be able to tell you very accurately, as long as he understands what you are trying to do, and understands that you wish a totally honest answer. In fact, all transceiver owners should run this test occasionally.

Should you be unable to get your transceiver to zero accurately, probably due to a monitor signal with no relationship to the transmit offset, a completely different approach is to fabricate an outboard audio oscillator, with a clean sine-wave output and a frequency precisely equal to that of the transmitter offset differential. The tone is then switched on at need and mixed into the receiver audio line. The operator then varies the received frequency until the signal he desires to zero is identical to the introduced tone, thus in fact zeroing his transmit frequency. A nice variation is to pulse the tone to give it a distinctive audio tag.

Whatever equipment you use, it is important that you become proficient and comfortable in fast, accurate and reliable zero-beating, as it is one of the most important keys to CW DX chasing.

Another area that may elude the new DX'er is the field of beam headings. If you own a rotatable antenna such as a Yagi or a Quad, do you wonder which direction to aim your antenna to work a certain DX country? Which direction is correct for India, or for the Cocos-Keeling islands, off the coast of Australia?

The answers to these questions are the great circle bearings for your location to those places. You probably covered great circle routes in school. You remember the situation, where your teacher drew a piece of string across a globe from one location to another on the globe, and demonstrated that the string represented the shortest path between the two points. If you don't remember, try it for yourself now. For example, run a piece of string from Chicago to London. Note that the compass direction at any point on the string is different from that a little further along on the string.

Radio signals are like the string; they always seek the shortest line between points. You will note that the string connecting Chicago to London on your globe leaves Chicago in a north-east direction; that, then, is the correct bearing to aim your antenna if you are in Chicago and wish to QSO London on the short path.

There is a special type of map that is designed to aid you in this; it is called an alti-azimuthal projection map, or, more simply, a Great Circle map. These maps have two virtues that are of great assistance to the DX'er. The important one is that they show the correct great circle bearing from the center of the map to any point on the globe, and also the correct and undistorted distance.

Such maps are available to North American DX'ers from the Radio Amateur Callbook and from the American Radio Relay League. If you have room on your shack wall for only a single map, that of the ARRL would be the choice, as it carries more information. If you also want and have space for a conventional map of the world, the Callbook Great Circle (alti-azimuth) map will likely be the choice in conjunction with the Callbook world map, as these maps together are more complete and cleaner in appearance. Alti-azimuth maps suitable for amateurs of other countries are generally available through their national radio society.

The great circle map should be in easy view from your operating position. Once you get used to locating countries on them, they are very easy to use. Simply locate, on the map, the location you are trying to reach, note the direction it lies in from the center of the map, point your antenna in that direction, and you will be pointed the correct direction for the short path to that country.

You will notice that the above sentence references the short path. For every short path bearing, there is also a correct long path bearing. But, since radio signals always seek the shortest path, how can there be a long path?

To answer this, consider what happens when we point our antenna and transmit in a certain direction. If the band is open and conditions are good, our transmitted signal keeps going on and on until it reaches a point at exactly the opposite side of the globe. What happens then? Do the signals drop off the edge there, or head for outer space suddenly? No, of course not. They simply continue on, going on in the same direction, though they are now more than half way around the world. Once past the point on the opposite side of the globe from the transmitter, the signals are considered to be "long path" signals. By pointing your antenna in the opposite direction, you can hear - and work - stations via long path. Often some of the finest DX'ing is done via long path.

An excellent example of this from the States is working into the Mid-East. In a winter morning, on twenty meters, the most reliable path into the Mid-East is the long path, with our antennas aimed into the southwest. The signal goes out across the central and southwest Pacific, then continuing on a great circle bearing into the Indian Ocean and on into the Middle-east, having come perhaps seventeen thousand miles. Yet, in winter, this is the most reliable path, more so than the typical short path bearing that would seem to make more sense.

Long path propagation is learned mostly by experience; but any time that you hear a big pile-up on a rare one, and you are having trouble finding or copying the station, don't be afraid to rotate your antenna 180 degrees for a look. You may get a pleasant surprise. And that is how you attain the long path bearing; take the short path bearing, and add, or subtract, 180 degrees, as appropriate.

There is one other significant exception to the short path bearing rule, which is called crooked path. It will be discussed further in the chapter on propagation. But if you are in Chicago, and hear a TL8 calling CQ 40 dB over S-9, peaking from dead south, what you are hearing is not crooked-path propagation. Rather, what you are hearing is a Pirate!

5
Basic Pile-Up Technique

I had finished dinner and was reading the paper when the phone rang. My daughter answered it, and called me. "You'd better hurry, Dad. He says it's about a new country."

I was at the phone in a flash. A voice intoned in deliberate manner, "Vee Kay Nine Yankee Tango, fourteen oh two three, short path. OK?" I recognized my buddy Jim, N6RJ, who had moved from Chicago to southern California several years ago. We used to work for the same company, and both of us are devoted DX hunters.

"Right. Thanks, Jim." I hang up. With a one minute long distance rate, we can and do call tips back and forth cheaply. In any case, I'm racing down the stairs.

Within moments, I am swinging the antenna to the northwest, the short path bearing for the Cocos-Keeling islands, one of the rare ones I really need. My "DX'ers Tout

Sheet" said nothing about any VK9Y/Cocos activity; I wonder where he came from, or if he's real. In any case, even as I muse, I'm getting everything going; work them first and worry later, as the old Bard taught me. Transceiver on, linear on, headphones on, antenna coming around OK, bandswitch to twenty, transceiver moved to 14,023. I listen eagerly, wondering whether we are going to have the same propagation the west coast is having; perhaps they have a lock on the fellow. But if that were the case, Jim would have told me before he hung up; he must figure I have a real chance.

OK, let's see what's what on the frequency. There, I find him almost instantly, "QSL VIA VK6RU OK 73 ES GL PETE AR N5TP DE VK9YT SK". OK, not too fast, not too strong; a good S-5, though, solid copy. Yes, he should be workable.

As I listen, I zero the VK9. I wait; yes, there's N5TP back to him. Hmmm. N5TP is about 400 hertz above the VK9; quickly I bring move up and zero on N5TP; after all, that's where the VK9 is listening. I adjust the keyer speed to approximate the speeds of the VK9 and the N5 station. "...R FB PAUL QSL VIA VK6RU FOR SURE 73 DE N5TP SK" I poise ready, waiting just a moment to see if there is any competition.

Hah. I should be so lucky. I hear eight or ten stations growling their calls out on the frequency, calling the VK9. A disciplined group as well; none of them called until N5TP was finished, indicating better than average courtesy and patience. Two or three of the stations seem to be pretty well ensconced dead on N5TP's transmit frequency; I decide I'd better move my transmit frequency up a couple hundred more hertz and try to get clear of those fellows.

That one second pause before I start calling has turned out to be consistently one of my better tricks; it gives me a quick peek at the competition; and a chance to react when I don't like the look of the cards I got dealt. I start my call. Lemeseehere... That fellow wasn't exactly setting the world on fire with his speed; must be a fairly new operator; best give him three times to copy my call. Yeah, OK, let's see how I did. I open up the receiver.

Oops, he's already back to someone; he's already giving a report. That means he picked up the other fellow, whoever he is, on a one time or a two time call at the most, while I was giving my call three times. So much for that theory. Well, let's see who he is in QSO with; then I can learn some more. OK, he's working WØSR. Where's SR? Yes, I hear him on backscatter; OK, about 200 hertz above me. Well, at least I moved the in the right direction. Hmm. Think I'll set up shop about 300 hertz above the WØ. The VK9 seems to be listening higher...

OK, WØSR is signing clear. Ready, set... Let's wait a second again, and see what's happening. Yes, there's quite a bunch calling him now. My frequency seems fairly clear. WØSR was going a bit faster than I was; I ease my keyer speed up a bit, and stroke my call out on the gleaming Bencher paddle. OK, that's enough, open the receiver and let's see what we hath wrought.

Ah, he's coming back, "WØVX DE VK9YT R TNX..." Darn! Another Zero. OK, the VK9 is turning it back; let's have a look-see for where the WØ is, and see how close we were to WØVX when we called the VK9.

Yes, there he is, on backscatter. Hah; he's almost dead zero on the frequency I called on. Ohkay, let's set up about three hundred hertz above WØVX. Hope I manage to pull this one off pretty soon; the pile-up seems to be getting larger fast, and no wonder; a VK9Y isn't just your ordinary pretty face.

...OK, the VK9 has signed with WØVX, and VX is starting his last transmission. Discipline is starting to break down; several stations are calling the VK9 on his frequency, ignoring the fact that the VK9 is working a station nearly a kilohertz higher in frequency. OK, VX is clear; I pause a second. Again, the frequency I have chosen to call on seems almost clear, but I can hear a real zoo starting up at the edge of my audio passband, below my frequency, down close to the VK9's transmit frequency. I start my call; I don't need to adjust my keyer this time; ØVX was going almost exactly the same speed that I was. I finish my call, and listen carefully.

The QRM on the VK9's frequency is terrible now, and at first there is no way I can copy him. I can't even tell if he is transmitting for sure, though I think I hear his signal in there. I wait; yes, he definitely is transmitting, "QTH HR COCOS ISLAND COCOS..." Yup, someone's got him. There has to be at least a chance it's me. My call seemed to have been timed about right. "...PAUL QSL VK6RU VK6RU HW CPY W9KNI DE VK9YT AR KN".

Allright!!!! I got him! Wow! "R VK9YT DE W9KNI R

GE PAUL ES TNX FOR QSO ES NEW COUNTRY UR RST 569 569 HR NR CHICAGO CHICAGO ES NAME BOB BOB QSL SURE HW CPY? VK9YT DE W9KNI AR KN". I listen with bated breath. Yes, "R W9KNI DE VK9YT..." OK, check the clock before I forget, and get the time in the log right. Too often in the excitement of things I have forgotten to enter the log time, and had to take a wild guess at it later. I've since learned to always log in the starting time right away.

"...FIRST DAY QRV FROM HR WILL BE QRV 4 DAYS 4 DAYS OK BOB NOT KEEP U 73 ES GD LUCK QSL OK W9KNI DE VK9YT SK".

"R FB PAUL U WILL BE VERY POPULAR HI HI TNX VY MUCH AGN 73 ES GD LUCK AR VK9YT 73 DE W9KNI SK". I feel elated and drained at the same moment. Wow! A rare new country.! But I'd better call it in on the two meter net. "Vee Kay Nine Yankee Tango, Cocos Keeling Islands, that's Vee Kay Nine Yankee Tango, fourteen-oh-two-three, fourteen-oh-two-three; short path, he's about S-5. I just QSO'd him, he's listening up about one kilohertz, here's W9KNI".

"Hey, Bob, way to go. I'm on my way! From K9QVB."

"OK, John. Good luck; good hunting."

"W9KNI here's K9BG. What was that call again, Bob?"

"Hello Jerry. It's VK9YT, 14023."

"OK, Thanks, Bob. Is he real, and is he loud? I just came in the door, and my wife said that you called something in and sounded excited."

"Yeah, I am. It was a new one for me. I'm pretty sure he's real. He has a VK6 manager. The path is right, it looks about as good as it could right now. Says that he's going to be QRV four more days. You're going to need your linear, though."

"Right, thanks Bob. I'm going for him. Here's K9BG".

"OK, Jerry. He's listening up a bit, one or maybe one and a half kilohertz, I'd say. Good luck."

"Thanks."

I turn back to twenty meters. Wow, what a zoo! I guess my call did get some results; four or five local signals are calling. They've got their work cut out for them; a mob of stations is calling the VK9, mostly on his own frequency where he's not listening. But for me, I happily fill in the log data, verifying that I have the proper date in GMT time, being as it is a day later than local time. I never did get my RST, but no matter. An exchange of signal reports, contrary to popular opinion, is not a necessary part of a QSO. I'll read it off the QSL when it comes.

I move up to 14,070, and swing the antenna down a bit from Northwest to straight West, and call N6RJ, my pal Jim who phoned me. 070 is our standard frequency, and I hope he is there. He comes right back, and starts congratulating me...

Pile-ups can provide some of the most exciting moments in amateur radio; few things in life will ever be sweeter than hearing your own call sign coming back out of a horde of stations desperately calling the prey; especially when it's an all time new country. But, clearly, when hundreds of stations are calling a rare one, and only a few are going to make it through, a pile-up equally is guaranteed to end as a failure and a frustration for the bulk of the people in the pile-up.

The observant student DX'er will discover that certain stations always seem to be able to get through the pile-ups easier and more quickly than the rest, be it in a contest or otherwise. And, more often than not, the successful stations do NOT have the loudest signals on the band. There are definitely tricks to the trade, and many of them will be taught and explained here.

But not all! There are techniques that can only work once in a given situation, and cannot be used again. Since the author hopes that a number of amateurs will be reading this book, they are best left unpublished. If you learn the methodology of this book, you will learn them for yourself one day; but in the meantime don't even ask! Also, there are tricks which are patently dishonest, and there will be no further mention of such here.

The best and most reliable approach to any pile-up is to already be in the log of the rare DX station! This statement seems simplistic; yet it is the point of much of the earlier pages of this book. Pile-ups usually start when lots of people hear a DX station being worked by a loud near-by station. But remember, the DX station got his first QSO of the operating session by calling CQ, or perhaps calling some one else calling CQ.

Also, it is very obvious upon inspection that your best

THE COMPLETE DX'ER   43

chance of working a new one is when you are the only one calling him. Your next best chance, all things being equal of course, is when there is only yourself and one other station calling him, etc., etc. Being the only station to respond to a QRZ or CQ of a rare station is far and away the best shot, however, since with no one else calling you don't have to worry about some potential competitor having a propagation advantage over you. W9's and WØ's will especially understand this! In any case, this is why careful tuning is always the best way to work the new one, rather than taking your chances in a pile-up.

But, pile-ups remain a fact of life for the DX'er. You can't be everywhere on the bands at once, and you doubtless have other duties which must be discharged in your day to day life. And, when there is a pile-up for one you need, you have to play the game.

One of the curious aspects of pile-ups for the skillful DX'er is that the better the operator at the other end the more difficult it can be to work him quickly. Fortunately, the corollary is that the better operator is the surer to work everyone. This sounds inconsistent, but there are reasons. The really skillful operator at the other end is consistent enough in his style of operating that everyone quickly learns his operating pattern, and many stations, rather than a select few, will be calling him where he is listening.

There is one variation of that, though, and that is the operator whose operating procedure is so completely random that it is virtually impossible to guess where he will be listening for his next QSO. Unless you get lucky, either operating pattern can make for a long night. Fortunately, this situation generally occurs on planned DX'peditions, and you can be reasonably sure that sooner or later you will prevail. Even if it is not a DX'pedition situation, you can keep hope; good operators have only gotten that way by being active.

Most operators, and this certainly includes most DX stations, operate in some reasonably consistent pattern, and the key to cracking a pile-up is to discover this pattern, and to adjust your transmit frequency and timing accordingly. Whenever you encounter a pile-up situation over a rare new one you need, you want to get your QSO as early as possible. Pileups, especially on non-contest weekends, have a rather nasty habit of growing enormously in a very few minutes in this age of two meter spotting nets, packet nets, buddy systems and cheap long distance calls, not to mention a considerable growth over the last few years in new licensees and new DX'ers.

There are several quick questions which must be answered when you come across a new one in a pile-up situation. Where is the station listening? Is he accepting tail-end calls? Does he move his receive frequency after every QSO? How long does he take to pick up a call? How fast (on CW) are the stations he works? The answers to these questions help you formalize your strategy for the situation.

Another aspect of a pile-up is to understand that pile-ups tend to be very fluid, dynamic situations, where the rules keep changing, as determined by the final arbiter, the DX station. And never forget, he has two penalties that he can impose, one individual, one collective. He can refuse to work you if you annoy him with your tactics, (or even worse - work you but NOT log you) or he can turn the rig off if too many people offend him.

In a dynamic situation where you have to keep moving, remember that there is no guarantee of success, ever. But, as things develop, if you keep your calls to the right length, and consistently are calling on a frequency near that of the successful station, you are bound to get the QSO if the band and the DX station's patience hold up.

When you encounter a pile-up, often the first problem you must face is to determine who everyone is calling, and where that station is transmitting. There are usually a number of clues that will be very helpful. If all the stations calling the DX station are rather tightly bunched together, chances are good he is working stations close to his own frequency, and you will likely find him in the middle of the mess.

If the stations calling are spread out over three kilohertz or more, that is generally an indication that the DX station is listening off his own frequency. Usually, some of the stations will be calling the DX on his frequency as well, accompanied by several policemen frantically trying to get them to move. In such cases the traffic cops themselves are the best beacon of the DX station's frequency, usually with the DX station unfortunately dead zero-beat beneath the loudest of the traffic cops!

Most DX stations listening off of their own frequency

listen higher than their transmit frequency, so you should listen below the pile-up first in hopes of finding the DX station. While this rule is not infallible, it is correct the great majority of the time.

Often, finding the DX station and getting his call may be two different things, with many stations signing their calls once every four or five QSOs, or even less often. Sometimes you can get lucky and find some one QSO'ing the DX station giving the call of the rare one, or even using the DX station's call while calling him in the pile-up. And, sometimes, finding the DX station can be made more difficult, especially when you lack his call sign, because you don't know what path to turn your antenna to while you look for the DX station.

Here, a bit of knowledge of propagation can be rather helpful. You should point your antenna, if it is rotatable, towards the best open path the band is carrying. If you have been carefully tuning the band before finding the pile-up, you have a pretty good idea what paths are open, and can check each of the likely ones. Sometimes, it is worth switching to a simple antenna, like a wire antenna for the low bands that will be largely omni-directional on the higher bands, until you have found and identified the DX station. This is especially true when the pile-up is huge, indicating a likelihood that the DX station has a decent signal.

And, of course, once you find out who the DX station is, you will often find that you have already worked the country. You may even occasionally find that the subject of the pile-up is a goody that you worked an hour before on his CQ, and now you spend twenty minutes trying to identify what all the excitement is about! Which is a good argument in this day of digital readouts to record the exact frequency of a station worked.

Having found the DX station, and recognizing that you need him, or while you wait for him to sign his call to see if you need him, you should be looking for the stations he is working so you can be ready to call once you have identified him. One of those simple truisms that seems to escape some of our brethren on the bands is that the frequency the DX is listening on is the transmit frequency of the station he is in QSO with.

This being the case, remember the call of the station the DX station is working, then, while still listening to the DX station, press the synch button (or the A=B button) on your transceiver, or switch to the outboard VFO if that is what you have, then quickly tune the second VFO to where you think the DX station might be listening, (usually higher, and usually in or above the pile-up) and hunt for the station the DX is in QSO with. If you have listened for a minute or so, you know roughly how long it is taking the DX station to process an exchange, so you have a pretty good idea of how much time you have to look for the station in QSO before switching back to the DX station's frequency to pick up the call of the next station he is QSO'ing. When you find the station the DX is working, you have found the frequency that the DX station is listening to.

Sometimes it may be several QSOs before you find the station the DX is working, and there are times when you never have much luck looking, due to poor backscatter, very wide frequency splits, etc. But you must do the best that you can at this; even the occasional spotting of a station the DX is working gives you more insight about what is really happening than you will get any other way.

In any case, once you have found the station in QSO with the DX, try moving your transmit frequency dead onto that station's frequency for a call when the QSO is finished. The reason you call exactly on the other station's frequency is because the majority of DX'ers seem incapable of accurately zero beating a frequency, making chances good that frequency will be fairly clear of QRM. Also, that frequency is an especially valuable one for the moment, in that the DX station is provably listening to it. However, this strategy is strictly a short term one, until you come up with a more complete strategy based on your observations of the DX station's actions.

You are now prepared to make your first call when the DX station has finished his current QSO. But, you must be very careful to keep your call short enough that you hear the DX station give out the call of the next station that he works. With luck, it may well be you! Failing to be so lucky, however, it is very important that you find the next station the DX works, and see how much, and in which direction, his transmit frequency varies from your own.

So, you called, and the DX station came back to another station. Find that other station. Is his frequency up 300 hertz from yours? If so, that is a strong suggestion that the DX station is listening up higher around 300 hertz after each QSO, so you should move your transmit frequency to be 300 hertz above the current station in QSO. If you fail to work the DX again, see what has happened after the next QSO.

Now, you begin a pattern of calling, then listening, and, when the DX comes back to another station, see where the other station is in terms of frequency, figuring out the the DX operator's pattern. Perhaps the DX station moved up 300 hertz again, and worked someone you were almost exactly dead zero with. OK, so you were out-gunned; move up 300 hertz again, and try again. Perhaps the DX station didn't move at all. That would suggest that he moves the receiver only if QRM gets too fierce on the last listening frequency. Your tactic then is to zero the last station, and pause a brief second before calling. If you are pretty well in the clear, stay where you are. If there are a number of stations calling on your transmit frequency, quickly move up about 300 hertz, and call there.

Does the DX station move his receiver up 300 hertz after each QSO? Or up 300 hertz twice, then down a kilohertz, then back up again? A common ploy of DX stations that keep moving their receivers higher up the band is that after a certain amount of stepping higher, they move back down to a frequency near their transmit frequency, and start the whole cycle again, moving higher with each QSO. This is especially true of DX stations who only have an RIT facility for listening off frequency, rather than an outboard VFO or separate receiver. Where you run into this situation, get some idea of what their high frequency limit is, and be prepared to call down at the other end of their indicated tuning range when they approach the high end.

A technique used by a few DX stations that is unusual is to maintain two pile-ups at the same time, one above their transmit frequency, and one below! In most cases, the DX'ers in one pile-up are not aware at all that there is a second one, especially the lower one. Almost invariably, the lower one is smaller. But, if the DX station is switching back and forth, you must switch as well with your calls, though your chances of success are almost always better in the lower one, but only when he is listening there.

Virtually every DX station has some sort of pattern, a pattern designed to help him find a station to work who is not

being obliterated by QRM from other callers. Learning that pattern is the key to success in a pile-up, and the quicker you comprehend it the faster you will get your QSO.

Every QSO the DX station makes while you are chasing him should give you more information on his pattern, sharpening your perception of it, and enabling you to more accurately place and time your calls. And that is your goal - to be the station in the path of the DX station's tuning, calling on a frequency clear enough that he can copy your call. When you get into the situation where each call you make is just about dead on the frequency of the successful station out of a pile-up, it should only be a matter of time before you are in the log, even though your signal may be many dB down from some of the big boys.

And, even if your quest for a particular rare one in a pile-up situation ends unsuccessfully, due to propagation failure or the DX station going off the air, what you have learned about his operating habits and patterns will very likely prove of value later - after all, you may be in a pile-up again for him tomorrow, and advance knowledge of his operating pattern will be a major asset.

6
Intermediate Listening

By now you have undoubtedly started testing the techniques learned so far from this book, and have begun listening to the bands in a more intensive manner. And, hopefully, you have already put a few new ones in the logbook. You have also learned that intensive tuning is pretty slow going. Now, we are going to try to learn how to tune faster and more effectively, yet with little risk of missing anything important.

As you recall, the main objective of tuning is to not overlook a signal, often rather weak, from a new country, nor a QSO in which a louder local station is in contact with a new one. All of this translates into not letting a new country get by your ear. Fortunately, there are a number of clues to watch for that quickly will let the experienced operator skip over the bulk of the signals that he hears. And, there are a number of optimization procedures to practice in your tuning that can speed up your coverage of an open band.

Suppose, for example, that conditions really stink, and twenty meters is the only band open for DX. In conditions like that, the intensive tuning pattern we learned earlier is ideal, because most or all DX signals are going to be pretty weak. And in truth, if the band is open to an area where you need a few countries, poor conditions can actually be more favorable for you, because the DX stations are more likely to appear on the only open band. Further, potential competition is likely to be reduced, since many stateside operators will be discouraged by the poor conditions.

But let's take a look at the alternative, a day when conditions are super, and three or even four DX bands are open, delivering good signals from far away places. Now, efficient tuning of the bands is essential, and a very slow pace of picking up each and every signal and having a careful look at it is more likely to be counter-productive. Clearly, a little more speed in covering each band would be helpful.

In light of these statements, you probably by now are wondering if all that stuff in the first chapter was bunk. Absolutely not. If you have acquired a habit of digging down into the noise and the QRM and dragging weak signals out to see where they're from on a regular basis, you already have gained a major tool for your DX'ing career. Now, we are going to sharpen that tool a bit further.

You will have noticed in your listening that DX signals often display characteristics that tend to identify them as DX. Since it is DX we are interested in, being able to recognize signals as being DX is a real plus. Once we have identified a signal as being from a DX station, we want to identify the call sign to see if it is one we need. So, always, when we are tuning, we try to identify every signal as being DX or not before we get a call sign.

Of course, on SSB, the accent of an operator is a dead give away that you are listening to an American, or to a DX station. Except that American operators are often sitting behind the mike at many a rare location. Again, some natives of other lands speak American English with a fluency that boggles the mind. Also, a very foreign sounding accent can be that of an operator settled into Los Angeles.

The CW operator lacks the benefit of an accent to help

determine if a station is DX or not, and must use other clues in making the determination. Some of these clues include polar flutter, auroral buzz, and subtle variations in frequency, very slight, due to doppler shifts on multiple paths. Of course, such clues can help determine whether an SSB station is DX as well, Yankee accent or no.

In some cases, tip-offs for the CW DX'er can include chirpy signals and hand sent CW, such signals often originating from those countries where modern gear and components are sparsely available at best, and getting a signal on the air at all, albeit chirpy, is a real feat.

In truth, we will never be able to guess correctly whether a station is DX or not one hundred percent of the time; there are too many quirks of propagation that can make a signal from one hundred miles away sound like rare DX for sure. On the other hand, there are times when conditions are such that signals from half way round the world come screaming in as though the fellow at the other end is across town. But that, unfortunately, is rare.

Obviously, the ability to distinguish DX from local stations can be useful, and it is an ability that the successful DX'er learns, indeed must learn. As you tune, guess whether each station you hear is DX or not. With a little practice the experienced DX'er can accurately identify a signal as DX or not DX ninety to ninety-five percent of the time. As you get better at this, you are sharpening another tool, and becoming more skilled and efficient at tuning.

Earlier, you were told to listen to non-DX stations as well, to determine if they are in QSO with a rare station. And, no doubt, you found that to be extremely time consuming. It is. But there are a number of clues to watch for that you can use to help speed up making that determination, with only a small chance of missing a goody.

Let's study some of those clues here. On SSB it's almost a give away. The SSB man, in QSO with a new one, will almost certainly make it obvious within a few seconds. "..Thanks, Jose, for the new country, my friend.." Not to mention the slow and deliberate speech used to speak to a DX station for whom English is a secondary language. QSO's with rare DX stations, on SSB or CW, are almost always brief; with minimal contact information exchanged. This may be regrettable in human terms, but it is a fact of life; probably partly because the DX'er in QSO with the new one doesn't want to be hoggish, and partly because he's so excited he isn't capable of carrying on rational conversation.

Also, the rare DX station often feels some pressure to pass out a lot of contacts, so he tries to work as many people as he can in the time available, and often does not encourage ragchewing. This gives you two important clues. If the QSO is particularly sparse, like "R TNX 579 579 BK" it's time to take a careful look at the situation. Or, if the station that you are listening to is making a lot of silly mistakes, try to find out if there's a reason; like he's in QSO with A51PN and is about to go into catatonic shock.

Another clue, on CW, can be the CW speed being used. On twenty meters, for example, most Americans like to QSO at twenty words per minute and more. Some of the DX stations, on the other hand, perhaps restricted to hand keys, or more limited in CW ability, will go slower. Thus, when you hear a W station going slowly and deliberately, there's a pretty good chance that he's into a QSO that he wants enough to slow down from his usual speed. Anytime you hear a slow fist outside the novice bands, give it a look.

Always watch for any mention of QSLs, with the exception of "QSL VIA BUREAU." If someone is saying, "QSL VIA K9AM" you'd best take notice, because K9AM is unlikely to be handling QSLs for a G station. The same is true when a station promises a QSL direct; it costs money , so there is probably a reason for it, like it's a new country for someone. Listen and see if it might be one for you too.

Context of a QSO can be helpful in other ways as well. If you hear someone saying that the temperature was 32 degrees Celsius, and it's the dead of winter out your window, you'd best hang around to see where it's so hot. Again, if a fellow ragchewing with his buddy starts complaining about mongooses killing the chickens, you just might want to stick in there long enough to see where it is the mongeese might be living. If there's talk of going QRT and heading for the sack, and it's breakfast time at your shack, you might want to pause long enough to copy a call.

Clearly, if you come across stations dealing out signal reports at a high rate of speed, you stay for identification. If you hear three or four stations saying, "QSY QSY" or "UP 5 UP 5" or "LID LID" or similar terms of endearment, or suggesting improbable physical activities on SSB, it's best to wait it out to see what the excitement is all about, making sure your linear is tuned up for that frequency while you wait. And, of course, you'll often run into pile-ups. But that is another subject, and will be covered further later.

If you are listening to a loud station you are pretty sure is a non-DX station, and he's going along at a nice comfortable rate, ragchewing or sticking to a fairly standard format, and there is no clue to suggest otherwise, he probably is not working anything likely to interest you, and you can with reasonable safely move on rather than hanging around to catch a callsign. Once in a while you are going to be wrong, and it will hurt a little, but far more often you will be right, and ahead of the game overall.

Now, let's go over some techniques for those times when conditions are really good. Let's pretend that we are operating at W9KNI again. It's a crisp late October morning...

The sun must have come up just a few minutes before my alarm goes off. I groan, and swing my feet out of bed as I sit up, searching for the slippers. I get out of bed quickly so I can turn off the alarm, so as not to wake my sleeping wife more than necessary. After all, it is Saturday, and she shouldn't have to get up. She's not the crazy one. I slip on the heavy terry bathrobe that she made me for my birthday to keep me warm in the cold basement shack, and slip out of the bedroom.

I stop off in the kitchen to put the coffee on, and head for the basement to fire up the gear. It is cold in the shack; I open the furnace vent to get a little warm air the next time that the furnace runs, turn on the gear, and return to the kitchen for my coffee. Soon, I'm back, settled in front of the rig, headphones firmly in place.

I swing the antenna from the South West, where I was having a last look at the Western Pacific last night before I went to bed, to the North-East, and turn the bandswitch to ten meters. Yes, immediately I start to hear signals; the band must be open. I start moving up the band. Tuning ten meters is pretty easy much of the time; every one you hear is either DX or someone QSO'ing DX. If conditions are any good, few local stations waste time working cross town or for that matter cross country.

Yes, the Europeans are in strong; listen to that YO3 for example, a nice clean S9. But it's unlikely I will find any

Europeans I need; I'm more interested in the possibilities of something from the Middle-East, where there are several countries I need. Maybe a JY from Jordan, a 9K2 from Kuwait, perhaps the LA5/A4X that the gang has been muttering about. Maybe even the legendary YI1BGD, from Iraq. The band is probably open to the Indian Ocean islands as well, such as the 3B8s and the FRs, which I already have, and the 3B6s and 5R8s, which I don't.

Fifteen meters is probably almost as good as ten meters, with a possibility of some trans-polar DX as well, but trans-polar DX is generally the preserve of twenty meters. Not that I'd refuse to work a new one if found on ten or fifteen, but my chances of finding a new one from that bearing are better on twenty when the band is open over the pole. Anyhow, fifteen will be the band of choice later in the morning, when ten meters is losing out...

The dilemma for the morning path DX'er with good conditions is whether to shoot for the short path on ten and fifteen, or the long path on twenty. Long path on twenty will undoubtedly be almost as good into the Middle East as ten is short path, but the competition will be tougher. Twenty has the additional advantage of probably better propagation into the FT islands of the far south Indian ocean.

Unfortunately, the short path on ten into the Middle-East and the Indian Ocean and the long path on twenty into the same areas open up at the same time and fade out at the same time, so the DX'er has to pay his money and make his choice. But, again, if ten is wide open, more DX and more DX'ers will opt for ten meters, regardless of what twenty meters is like. Probably the best ploy is to work ten meters on the first morning of a weekend, if it is nicely open, and try twenty long path the following morning. Also, the long path might still be open on twenty the next day even if conditions are slightly disturbed, cutting more into ten meter propagation than it will into twenty.

Ten meters has an inherent trap that even relatively experienced DX'ers easily fall into. When the European signals are loud and strong, the DX'er looks past them, trying to find the rarer and more elusive Middle East and Indian Ocean DX stations. But, after several hours the band closes to the Middle East, as darkness encroaches in those areas and wipes out their propagation terminal, yet the European signals will remain strong for several hours after the Middle East fades. The DX'er is fooled into thinking that the band is still open that far, since Middle East stations are uncommon enough that he fails to notice their absence.

That is the time that the clever DX'er searching for the Middle-East has switched down to fifteen meters; with any luck the hordes are still mesmerized by S-9 DL's on ten while you are logging the 9K2 on fifteen. Later, when fifteen too has lost it's path into the Middle East (And the Europeans are still coming in strong on fifteen) the sharp DX'er goes back to ten for a brief look around for Central and Southern Africans on ten meters; then heads to twenty meters in hopes of an early afternoon opening short path into the Middle-East again.

Of course, as we started out saying above, we are making the presumption that we are located in the American Midwest on an October morning, and the above strategy applies to those circumstances. They may well bear no resemblance to the conditions at your QTH. But, similar situations do exist for every DX'er and every QTH, and the best way to resolve conflicting propagation strategies is through the use of the kind of logic worked out above.

There are several other considerations that deserve attention in deciding what one's best strategy is. For one, the DX'er should keep track of what countries he needs, of course, and this can affect his tuning plans. For example, if he has Europe and the Middle-East pretty well worked out, but is weak in the Africans and the Pacific, an entirely different strategy would be called for, dictated by the need list and propagation.

You would sleep in late Saturday mornings, having been up 'till the wee hours the night before tuning twenty for those areas. You would start again in the late morning to check ten and fifteen, for the Africans first, and then the Pacific a little later as the sun rises in those parts of the world.

Another aspect you will want to consider is what bands you are most effective on. If your antennas are definitely below average, you probably will do better to spend more of your time on ten and fifteen meters than you will on twenty, where the competition is tougher, and lower antennas less effective, comparatively speaking. And of course, careful tuning is all the more essential if your antennas are not as good, since you want to work all the harder for that one-on-one shot at a new country. But, even if your antennas are not so hot, spend some time on twenty. It still can be a very productive hunting ground for the DX'er with the less well endowed signal; it just takes more work and time.

There are other tricks of the trade as well. One is the day and date. Consider, for example, that in the Moslem Middle-East countries, Friday has the same status as Sunday does in those countries where Christianity is the dominant religion. So, even though some of the stations of the Middle East are operated by Westerners, Friday is the day off, and the most likely time that some of the stations will be active during what would otherwise be working hours.

A useful book for serious DX'ers is an annual almanac, since they generally list the public holidays and the religious holidays of many countries, and the astute DX'er knows that his chances of finding a DX station on from a particular country are considerably enhanced on such days.

Another consideration, along the same lines, is time and date. For example, take a look at your world map pasted up over your station, and take note of the international date line out in the mid-Pacific. As you know, stations west of that line are a day later in calendar date. So, what is Friday evening for us is Saturday afternoon for them, and they may well be active, celebrating their weekend by waiting for you to answer their CQs. Thus, Friday evening is a great time to hunt for those lads. Sunday evening, on the other hand, clearly is not so good, because most of those chaps are back at work, bemoaning Monday.

In fact, local time at the other end is always an important factor to consider, for various reasons. For example, Middle-East stations tend to operate more during the morning and evening hours, since the heat of the day can make for very uncomfortable operating conditions in those parts where air conditioning is not necessarily a fact of life. Again, except for the dead of winter, conditions on any DX band are poor around noon local time for any particular station.

Rare DX stations for some reason don't generally respond the same way to clocks that determined DX hunters do; they go to bed at night, and sleep. If it's three AM in Kuwait, your chances of finding a 9K2 on the air tend to be rather slim. So if a 9K2 is the last one that you need that's on, you'd best take a look at a time conversion table. Your best chance to find a 9K2 would be on a Friday morning, either on ten meter short path, or twenty meter long path. But if it is the weekend of one of the World Cup Soccer games, forget Kuwait - they are rabid soccer fans, and have a fine highly competitive national team. No Kuwaiti amateur in his right mind would risk TVI - and his life - during one of those games!

As you can see, an important part of the game is knowing when to get on a band, and when to get off a band, and what days are best. Most of your tuning time will be spent not looking for a specific station, but rather tuning in general, always hunting. But you always want to try to do it in such a way that you are optimizing your chances of finding a new one, no matter what your country total, as long as there are stations on the air you need.

7
Intermediate Pile-Up Technique

    I was sitting at the operating table, the gear on, but I was not listening. Instead, I was tussling with the checking account, trying to figure out why the bank had more money and I had less. I had checked twenty meters earlier, and conditions seemed good, but it was the hours of "in-between," where the band is open but there's little of interest on. The Europeans have all gone to bed, and it's too early for the Middle-East to be up and out of bed. Of course, I could go try the Asian path on 15 meters, and as far as that goes, the Soviet Asian countries should be coming through on twenty short path, not to mention the VU's, 4S7's and the AP2 gang. But, if I don't figure out my checking account error, that two-element forty meter beam is a couple more months off, and I'd like to have it up this fall.

    But all my calculations are shoved aside when the two

meter rig speaks up, "Say, does anyone know what the pile-up on 14033 is about? Here's W9NUD"

My receiver is sitting on 14,025, so within a few seconds I've got the headphones in place and I'm on 14,033. I hear a few stations ending calls, then the frequency is quiet. The antenna is straight north; I decide to leave it there until I get things sorted out. No one is calling now; whoever the pile-up is over must be transmitting now. I listen very carefully. The stations I heard calling were sending fairly slowly, which is a helpful clue in seeking the DX station, suggesting as it does that the DX station is also sending slowly.

There's a station, not very loud, bit of a chirp, in QSO. Hmmm. Speed's about right, and he's near where the pile-up was. He's drifting a bit, too, going down in frequency. I can copy him OK, let's see who he is... "569 QTH HR ISTANBUL ISTANBUL NAME IS KAMEL KAMEL HW CPY K8MFO DE TA2BP KN" Hey! It's Turkey! That would certainly be a new one for me. A rare one, too, because amateur radio has only recently become legal in Turkey. Hope he's real; but right now I'd better be more concerned about getting a QSO than worrying about the legitimacy of his license. As the fabled Bard said, "Work them first and worry later."

Snap! The linear surge relay drops out, and the linear is up and running. I glance over at it; yes, it's still tuned for twenty meter CW, we're OK there. Oops, the antenna is still pointed straight north; better pull that around to 45 degrees; that isn't the exact bearing, but it's close enough. OK, we're ready; now, where's K8MFO? He's not in my receiver passband. I press the transceiver synch button, then start tuning, looking for 8MFO. I decide to try listening up one or two kilohertz.

Yes, there he is, about 034.5... "QSL SURE KAMEL ES TNX NEW COUNTRY 73 ES GUD DX TA2BP DE K8MFO SK"

I decide not to transmit my first call on MFO's frequency; even though I haven't really heard the pile-up properly yet, from what W9NUD said when he called it in on two meters it's obvious there is one, and that means that there will probably a number of stations on 8MFO's frequency. MFO is about one and a half kilohertz higher than the TA's transmit frequency, so it seems certain that the TA is listening split. I decide my best shot is to set up about 300 hertz higher than 8MFO's frequency. I set my transmit frequency there, then switch to the other VFO to copy the TA2 again.

Yes, there he is... "TNX DR DON QSL IS SURE 73 ES DX DE ISTANBUL K8MFO DE TA2BP SK QRZ? K" As he transmits I study his signal more carefully. Yes, he is drifting down, but not too badly. Slight chirp to the signal. Speed about 15 words per minute; not half bad for a hand key, which he is obviously using, and doing a fairly good job with. It's a heck of a lot better than my fist on a hand-key. I'd be lost without my Bencher paddle.

When he completes his transmission, I pause a second before I start my call, listening on my transmit frequency. I can hear the growl of a pile-up below my frequency where MFO made his QSO, but my own frequency seems reasonably clear. I seem to have guessed right in getting off MFO's frequency. Now let's see if my guess is going to do me any good. This could be a tough one; there's a lot of stations calling him, and why not? Turkey is a rare one, and this fellow has a decent signal. I call, "DE W9KNI W9KNI W9KNI AR"

I listen again. A few stations calling him on his frequency, making copy impossible, but I can tell he's transmitting. I wait, and one by one the stations on his frequency stop calling. I listen to see who he's in QSO with; OK, it's K9DX. He's in our local DX club too; I guess I wasn't the only one to respond to W9NUD's call on two meters. Well, at least it should be easy to find K9DX. He's always loud here.

When the TA2 gives it back to K9DX I move my receiver back up to my transmit frequency. Yup, sure enough, there's DX, about 300 hertz higher than I was calling. Since I set up three hundred hertz above MFO, and K9DX set up 300 higher than me, that means the TA moved up six hundred hertz after clearing with MFO. So, I move my transmit frequency up 600 hertz above DX's frequency, and wait to make another call...

The TA2 sends QRZ, and I call. Again, I give my call three times, and wait for the TA to come out of the QRM. After a moment, I realize that he still is not transmitting, so I call again, but only signing my call twice this time. I open the receiver, and this time I hear him right away; apparently the fellows calling on his frequency had finally waited too. OK, he's in QSO with K8EJ. I wait it out, OK; he turned it over, where's K8EJ transmitting? I move back up. Right, he is transmitting almost exactly on my frequency. Guess he's either luckier or louder than I am, probably both. Anyhow, I move my transmit frequency up another 600 hertz. This ought to be easy now, I've got the TA's pattern. I wait, then call again.

This time I have to wait for the QRM again, but I hear him in there. I can't hear who he's back to, but I wait. My timing wasn't bad, maybe it's me. There, he's turning it back over. Darn, some K5 got him, yes, K5LM. Drat, I thought I was almost in line. Let's see where K5LM is transmitting. I move the receiver back up to my transmit frequency. Nothing, not a darn thing. I look a bit higher. Still nothing. I wonder if K5LM knows he's got the TA. I go back down to see if the TA2 has started calling the K5 again. Nothing doing there either.

Wait, there's the TA coming back. But he started his transmission with a big "R". And he's definitely responding to K5LM's transmission. Something's funny here. I should certainly hear K5LM; that's an easy path from here. I wonder where I went wrong? And I have an immediate problem; where should I call? I'm completely lost. There, the TA is signing clear. In desperation, I move up another 600 hertz, and call. Maybe K5LM isn't in skip distance, and I just didn't hear his backscatter signal. Deep down inside, I know better, but I don't know what else to do.

I wait, knowing I really don't have a chance this time; something is wrong. Ah, there, he's coming back. Hmm, that will help, W9NUD nailed him. He's only about five miles away, so I should have no trouble finding him. The TA completes his transmission, and I go looking for W9NUD. First, I check around my own frequency. Nothing. I look higher. Still nothing. This is getting ridiculous; I know I can hear NUD. I start tuning lower. Hah! There he is. About two kilohertz below the TA's frequency, and definitely in QSO.

Now that I know where NUD is transmitting, I have to decide where to transmit. Is the TA moving higher or lower, now that he has shifted to the other side of his transmit frequency. I decide to call lower; if he's using a transceiver, he is probably using the RIT control to move around with, and is staying away from his own frequency as a smart DX operator will always do when coping with a pile-up. So, probably what happened is that when he was on the high side of his frequency, and tuning higher every time, he ran out of RIT range, and elected to go to the low side, where 'ole K5LM was lying in the weeds waiting for him. And, W9NUD figured out what was happening, a lot faster than I did.

I set my transmit frequency about 300 hertz lower than W9NUD. With a little luck, the pile-up below the TA will be rather smaller than the one above him. At least I hope so. I'd like to think I got smarter quicker! Even if I am lucky that NUD lives only five miles away from me. And there's little hope for those stations calling on the TA's transmit frequency; he is clearly astute enough not to give any of those fellows a chance.

The TA signs with W9NUD. I sign my call twice on my new frequency just below NUD's. Hey! I got him! Hot Dog! Wow! Hooray! "W9KNI W9KNI DE TA2BP R TNX OM FOR QSO UR RST 569 569 HR QTH ISTANBUL ISTANBUL ES NAME KAMEL KAMEL HW CPY? W9KNI DE TA2BP KN"

"R TA2BP DE W9KNI R FB DR KAMEL ES TNX QSO ES NEW COUNTRY UR RST 569 569 ALSO HR NR CHICAGO NR CHICAGO ES NAME BOB BOB HW COPY DR KAMEL? TA2BP DE W9KNI KN"

"R W9KNI DE TA2BP FB OM BOB RIG HR HOMEBREW TRANSCEIVER 150 WATTS ES LW ANT PSE UR QSL VIA TRAC BURO MUST QRT NW 73 ES TNX DR BOB W9KNI DE TA2BP SK QRT CL"

Whew! That's cutting it a touch fine. If I hadn't found W9NUD, I wouldn't have gotten that one. Only during NUD's QSO did I finally figure out what was happening. If it had taken one more QSO, someone else would be logging the TA2, and I'd be one of the unhappy DX'ers who didn't make it when the TA2 went QRT.

I note the time in the log book, being sure I get the GMT time correctly, then pick up the microphone on the two meter rig for a little post-mortem with W9NUD, and to thank him for calling it in. And I want to know how he knew to go looking below. Was he lucky or smart? In my case, I know the answer. I was lucky! But that's part of DX'ing too, and especially in the pile-ups.

And that is how it goes, very often. Especially in the pile-ups, you are dealing with incomplete data in a fluid, dynamic situation where you can only respond, trying to guess what the DX station is going to do next, and at the same time trying to outguess the competition.

When you are in a pile-up situation, you want to make your signal as attractive as possible to the DX station. This would seem obvious on the face of it, but plenty of DX'ers never seem to get the message. On CW, your keying should be flawless, and pleasant to listen to, and sent at the speed the DX station appears to prefer, as evidenced by the speed of the stations he is working.

On SSB, especially when calling a station for whom English is clearly a second language, you should call in slow, firm and clear phonetics. Clever phonetics using words unfamiliar to a DX operator are almost sure to fail; if your call sign suffix is JWU, use "Japan, Whiskey, Uniform," rather than "Johnny's Wet Underwear," for example.

Always, in a pile-up situation, you must make every possible effort to see that your signal is in the clear. In a big CW pile-up it is not reasonable to expect to be the only signal inside a 500 hertz bandpass, but on the other hand you don't want to be zero-beat with four other stations either. On SSB, you don't want to be sharing a frequency with twenty other stations. None of you will likely get the QSO from that frequency. DX stations want a reasonably clear frequency to listen to; that is why they keep tuning their receivers.

Your tactics in a pile-up should change to some extent depending on the size of the pile-up, and your evaluation of the likelihood of success based on normal tactics. For example, if you hear a very rare one on with a thousand stations locked in heated combat for his attention, and you are running a hundred watts to a vertical, you are going to have to be rather more innovative than standard pile-up practice suggests if you are to have a chance to log him.

In fact, the very first question you should ask yourself is whether you have a chance at all? The question is well worth asking. Let's analyze a few different cases to demonstrate this. First, we'll look at a similar situation. You are running a hundred watts and a ground plane vertical, and the ZD9 which you need for a new one is on. With hundreds of DX'ers calling him. The ZD9 is a good 569 at your QTH, going rather slowly, and does not appear to be moving his receive frequency much at all. You have 156 countries, and you've never even heard a ZD9 before. What do you do?

You listen for five minutes, taking careful note of his fist and his pattern. You listen to people QSO him, and you note that they are all the loudest people in the pile-up. You memorize the characteristics of his signal; then you QSY. You haven't got a prayer of working him under those circumstances, and on the other hand with him attracting all the hot-shots on the band you have an excellent opportunity of finding another rare DX station calling CQ, all alone and lonely, that you need every bit as bad. Remember, being a successful DX'er is the art of the possible.

OK, now we change one aspect of the above situation. You are a hot shot DX man yourself, and your hundred watts and ground plane have 296 countries worked and confirmed. (If you are a newcomer to DX'ing, be assured that such a score is possible with such gear. For example, the author worked 210 countries in 11 months similarly equipped.) Now, hopeless though it seems, you stand and fight it out. Why? For self punishment? No. Even though you have a much higher score, and obviously know the tricks, you still have two chances of working this ZD9, slim and none.

But, with 296 hanging on the wall, the ZD9 patently represents the only game in town. Even though you probably aren't going to work him, you want to learn everything you can about him, and how he works through a pile-up, because you're going to be chasing this one for days, weeks or months until you get him.

OK, now we change it again. You have 137 countries. Conditions are pretty poor, and this shot into the South Atlantic is probably the only chance you are going to have at a new country tonight. And the ZD9 seems to be a pretty sharp operator. He's moving his receiver a lot, keeping away from the big concentrations of callers. His fist is snappy, and his signal good. This time you stand and fight it out on the line of scrimmage. You certainly don't have a sure thing on your hands, but with a little clever operating, you have at least a fighting chance at a QSO, and besides, you'll probably learn something, win or lose.

So, you've made the decision as to whether you are going to stick around and play the game or not. If you aren't going to stay with the pile-up but instead play the odds and tune for another new one, your next move is clear cut - turn the knob. But if you are going to stay, it's time to take stock. You must decide whether your signal is likely to be as strong as most of the people in the pile-up into the DX station's receiver, or rather weaker, or somewhere in between, including allowance for propagation and geography?

If you have one of the stronger signals, your tactics should be to stay rather conservative, and play the short odds. You keep your transmit frequency close to that of the last station to QSO the DX, as long as that seems consistent with the DX station's pattern, knowing that if you keep calling pretty well on the right frequency your number is just about bound to come up sooner or later.

But no, you say, your signal is the kind that keeps linear builders putting up new factories. OK, how much weaker are you than the pile-up leaders? If you are not more than a couple S units down from most of the stations getting QSOs, i.e. they are from the same area of the country you are, and you are not giving up more than 10 dB in antenna and power, you still stay fairly conservative and careful as well, though not quite as much.

For example, if the DX station is tuning about 300 hertz higher with each QSO, you try to stay about 500 hertz higher than the last QSO instead of 300. Your hope here is that a bunch of the heavies on frequency will figure out the great secret of life at the same time, thereby QRM'ing each other so badly that the DX station will QSY up further than normal, only to find you bleating out your heart and callsign in solitary splendor.

Also, if all the heavies are getting through by signing their calls two times, you always sign yours three times, hoping that the QRM will open at the last moment allowing you to slip in that one clear call. This technique was Alfred Lord Tennyson's best shot.

You keep nibbling away at the fringes. And, while doing so, you keep studying the DX station, and the tactics of those successful in getting a QSO. You try to figure out the exact timing the DX station likes. You try to anticipate his every move, all the while maneuvering for the break you need. This may sound silly at first, but you will find eventually that there are breaks, lots of them, in most pile-up situations, where a skillful call, perfectly placed and perfectly timed will get through, even though you are several S units below the madding hordes. And, when you do snag the really rare one and get a 569 report and all he has been handing out for the last hour is 589 and 599 reports, your shirt buttons are in real danger, and you know you are a DX'er.

But no, you say, you are running real QRP, and are way down in signal strength in terms of the other denizens of the madhouse. If the pile-up is really large, frankly speaking your chances are virtually nil, unless you are a close friend of the DX station, and he knows your call extremely well. Then your chances are greatly enhanced, clear up to being very poor. The biggest problem is that the stronger signals are going to so totally capture the AVC system of the DX station's receiver that practically speaking your signal is going to get wiped out. If the pile-up is a large and avid one, your best bet is definitely to go tuning off, looking for another rare one. Your chances of snagging one elsewhere are enhanced by the presence of the pile-up, because most of the DX'ers active on the band are caught up in it, leaving the rest of the band to you.

But, if you really want to have a go at the DX station in the pile-up, there are still things that you can try. For one thing, if the DX station is listening five kilohertz up, you can try calling three kilohertz BELOW his frequency. Occasionally, some DX stations will have a look below, and you might get lucky. But don't count on it.

There are two other special pile-up techniques; the delayed call and the tail-end. These will be covered in detail in a later chapter. In the mean time, keep your frequency clear and your calls short, and good hunting.

8 Advanced Listening

If you started as a beginner taking the lesson approach in this book, by now your listening skills should be considerably enhanced from your starting point. You should have the ability to "work" a band just as a politician works a crowd, able to ferret out the DX, and to separate the wheat from the chaff. You should be able to copy weak signals out of heavy QRM and QRN, at low speed or high. And these improved skills have no doubt contributed to your country total. You have probably even discovered that this type of listening is a lot of fun, with always the possibility of a new country at the next kilohertz.

But the approach taken so far has been somewhat a random one. Since intensive tuning takes a good deal of time, we must consider ways to make optimum use of the time available for it, taking into account our other obligations to family, job, school etc. Also, we should note that some of the rarer DX stations are exceptionally cagey, and, disliking pile-ups, are rarely heard on the bands. This does not necessarily mean they are not active though, so we will discuss a few techniques for unearthing them.

First, let's discuss propagation a bit further, and its relationship to tuning for the new countries. As noted in an earlier chapter, the majority of DX'ers seem to take the highest band open, turn their antennas to the loudest DX path, and listen. And, if you have fifty countries, that is a good plan. With fifty countries, it is fairly safe to say that you should be able to work a new country on almost any path almost any day.

On the other hand, if you have 185 countries, things don't work out quite that neatly. With 185 countries on the wall, you have obviously worked just about all the common garden variety DX around. Now, you are more interested in sniffing out the rare ones that choose to lurk away from mainstream DX routes.

An important source of DX often is the secondary propagation paths. These paths are mostly to be found on fifteen and twenty meters, and more often in the spring and fall months, but they exist year round. For example, a good secondary path from the American Midwest occurs in the summer months. The primary band path in the late afternoon, around 2100 Zulu, is into Europe and the Middle East. But, if conditions are reasonably good, there is often a fine path into Japan and south east Asia at the same time. Admittedly, signals are several S units below strengths on the North Atlantic path, but regardless they are definitely Q5. And they often display some rather nice DX, such as YBs, 9Vs, DUs, 9M6s, V85s and VS6s etc.

Often, secondary paths can include good long path openings, some of them virtually as reliable as their short-path counterparts. An example of this, again appropriate for the American Midwest, is the path into southern Africa. In the morning hours after sunrise, this can be an excellent path, with antennas west and signal strengths on twenty meters that have to be heard to be believed. This is an excellent path for FRs and FTs, 5R8s, 3Bs, S79s etc.

Hand in hand with that path often is the long path into the Middle-East, and occasionally a shot at some of the really good (for which read "rare") DX. There are more paths like this for the Midwest, but similar paths are regular occurrences from any QTH. Often, if in doubt, an easy way to find them is to point your antenna in one direction or another off the usual direction for the frequency and time and let out a good CQ DX, especially if you have a linear. You may get a rather pleasant surprise. Once you have determined the existence of such a path, be sure to check it out regularly. Such paths often will have DX glad to work an occasional W station as long as they don't instantly become the subject of a howling pile-up.

The point to remember is that secondary paths often provide a chance to work DX stations at times that are optimum for them to be active. For example, an early morning long path opening into the Middle-east catches them in late afternoon their time, which is a convenient operating hour for them, so that they are more likely to be on. You can't work a station that is turned off with its owner asleep in bed.

The likely operating times of DX stations we need clearly is an important consideration. Earlier, we talked about weekends, public holidays, etc. But, equally important is being on at a time that the DX stations are likely to be on. This means, for example, that to work the Pacific DX islands reliably from the American Midwest it is necessary to stay up till the very late hours of the night, or to rise at truly terrible hours, such as 3 AM local time, to snag elusive little goodies like a VKØ on Macquarie.

Remember, DX stations often are not interested in country totals, nor are they particularly interested in getting up at funny hours and skipping meals just so they can QSO another thousand Americans. If they want to work Americans, it will be at their convenience and pleasure, not yours.

Some DX stations, though, are interested in working a rare one. And, often, what is rare for them is not necessarily rare for you. For example, an XE in Mexico is considered an excellent catch in some parts of the world, and an XE with a good strong signal can raise some fine pile-ups of DX stations, as indeed can most Central American and Caribbean stations.

This, in turn, offers a good chance for the patient DX'er to snag a goodie out of the pile-up. The technique is simple. Find a "local" DX station doing his thing with a pile-up of European or Asian DX stations. Start listening to the pile-up itself carefully. Chances are surprisingly good that you will find a goodie or two in there calling.

But, like the 7X station in the earlier chapter, working them requires finesse. To set your transmit frequency dead on the frequency of the DX station's quarry and give him a call is just about guaranteed to fail, not to mention earning you the enmity of the pile-up at large. Rather, you must work very carefully.

There are several possible ways to get your quarry, depending, in large part, on what the other DX station is doing. For example, if the XE is, let's say, slow and deliberate, giving out his name, QTH, power level, weather etc. with every QSO, and the V85 you are trying to snatch out appears to be a rather competent chap, you might try setting up your VFO about 300 hertz above the XE's frequency, and, just after the XE has started a transmission to some one else, with the V85 waiting patiently, blast out a fast "5CM DE W9KNI W9KNI 569 BK".

In this way, you are telling V85CM that you want a QSO, that you do not wish to mess up his chance with the XE, and that you won't attract further attention to his exotic callsign, since you left out the V85 part. With any luck, and if you did it politely, you may get a response, "W9KNI R 579 QSL 73 SK". That's all, but that's all you need. You would respond, "R TU 73 .." Clearly, a very minimal QSO, yet one in which the nuances exchanged equal the actual QSO information and a new country in the log. And that's what it's all about.

On the other hand, let's say that the XE is a pretty snappy operator himself, whereas the DX station you are trying to grab does not appear quite so sharp. There are several possible approaches to this situation. One is to wait to see if the XE finds the DX station himself and QSOs him, then leaving you the chance to call the DX station slightly off the XE's frequency. asking the DX station to QSY down five, perhaps.

Another is to call the XE yourself, trying to secure a quick QSO, and informing him that the rare DX is calling him. It may well be that the rare DX is stronger at your QTH, and the XE will appreciate the information. Doubtless, you will ingratiate yourself with the rare DX station as well. The trick to this is to turn your antenna towards the XE to give your signal enough punch to get the QSO started; then, once the QSO with the XE is established, immediately turn your yagi back to the rare DX you are hunting. The XE will still be able to easily copy you, but you want to make sure the V85 hears you telling the XE that the V85 is QRX. Then, as soon as the XE is finishing with the V85, you ask him to QSY off for your QSO. This scheme is hardly guaranteed to be a sure thing, but skillfully done it can work the majority of the time. And nothing ventured, nothing gained.

The other approach is to sit it out, waiting for the XE to QRT, should he not QSO your quarry. When the XE closes down, it can be a bit tricky, since the station you are interested in will be disappointed in not QSO'ing the XE. Also, he will be hoping somehow that the XE will still hear his call and come back to him. Your best bet in this situation is to be sure that you are at least several hundred hertz off the XE's frequency, then call the DX, usually without using his entire call, with a message that you are QRX. Like, "5CM DE W9KNI W9KNI QRX HR".

Again, we are telling the DX station that we do not propose to draw attention to him, and that we understand his wish to QSO the XE first. At this point, presumably, the V85 knows the XE is probably QRT, and hopefully will appreciate your politeness enough to give you the QSO.

Careful tuning, as you will have learned by now, will be constantly turning up rarish DX stations engaged in their own operating situations, such as in QSO with a pal, chasing their own DX, calling a directional CQ for which you do not qualify, calling a station on schedule etc. Most of these stations can be QSO'ed, albeit briefly, but only if you adopt an extremely courteous approach. After all, they are interested in such circumstances in something other than a QSO with you.

The approach you need to take in these situations is almost always the same; call them several hundred hertz off the frequency you think they are listening to, at a time you are sure there is no one else transmitting information they would rather be copying, and always only after a pause that shows them you understand they are busy elsewhere.

Similar tactics can be used on SSB. For example, if you hear the V85 calling "CQ South America," delay several seconds after he has completed his call, and, when you are sure he is getting no response form a South American, make a brief call, If that call goes apparently un-noticed, simply wait. If the V85 takes no apparent notice of you, and calls "CQ South America" again, wait. After he has made a SECOND CQ call to South America with no response, repeat your call again,

"5CM, here's W9KNI QRX for you, Old Man, W9KNI QRX."

Never make a big thing of their call under such circumstances, as it will be sure to attract a pile-up when others hear you calling the station. Which is the last thing the V85 seems to want. Respect his wishes, and he may grant yours.

Another sneaky tactic that can help in such circumstances is to make such calls at reduced power. A DX station NOT looking for a W QSO may very well take more pity on a weaker signal than on one that blows his eardrums out, and may further figure that a quick report is less likely to lead to an unwanted pile-up of howling DX'ers. If several careful calls go unrewarded, you can always turn the linear on and try again. If you know your signal is likely to be very loud at the DX QTH, it will always be wise to reduce power in such one-on-one situations where you are not the station the DX seeks to work. An S-6 signal pleading for a QSO is more likely to be acknowledged by the DX station, whose intent is elsewhere, than one 20 dB over S-9. And, if you are the only one calling him, the power is otherwise not important.

Although you must keep QSO's thus snagged very short, you always want to insure that the DX station understands you will be after a QSL, to help see that he actually logs the QSO. Generally, something like TNX NEW COUNTRY or "thanks for the new one, and I'll be sending my QSL tomorrow" will do it, but be sure you let him know you will be QSL'ing the QSO. And get the QSL out as soon as possible via air - if he failed to log the QSO, perhaps he will remember you.

An old saying for the hunter is that to hunt for a tiger you have to think like a tiger. There is a natural corollary for the would be DX'er; if you want to work rare DX, you must think like rare DX. While this is something of the essence of pile-up strategy, it has other ramifications as well. Let's pretend for a moment that we are a rare DX station. Now, the reason we are considered rare is that we don't have much interest in working another five thousand W stations so that we can exchange reports and QSLs. We already know very well that our signal is capable of reaching any part of the world, and meaningless QSO's to prove it are of no interest to us.

Yet, we continue to be operational, because we do enjoy ham radio. The question becomes how do we stay active without always raising huge pile-ups? There are several tricks.

THE COMPLETE DX'ER

For one, we operate, when possible, at the times that the W operators are mostly in bed, or when propagation is not favorable for W stations. For another, we use a directional antenna, and try to minimize our signals to those areas where we will encounter pile-ups. We never call CQ, preferring to only answer those of the stations we want to QSO.

When we run schedules with our pals, we do it on funny frequencies, like 14110, or 21188. We use call signs only at the bare minimum, in fact often rather less than our licensing authorities prefer. We are not totally against working W stations, but we certainly don't need more wall paper.

Now, reverse roles again. We are the American DX'er interested in QSO'ing that elusive one. We know that there are DX stations active that we need like the fellow described above. In truth, both Africa and the Pacific are full of them, operators regularly active from countries that most DX'ers are prepared to swear have been off the air for years. But, having thought out the DX station's position, and being aware that he is not interested in pile-ups, we start searching the byways as well as the highways.

We search all the useable DX frequencies regularly, not just the bottom fifty kilohertz of the band, and we do it with the RF gain wide open so we can hear the weak ones, off the side of their antennas. We get up regularly at funny hours of the morning to have a look around, both to the West and to the East, especially on twenty meters. We listen to the ragchews, deciding whether there may be something good in one, especially if the band is uninteresting. We regularly check out the fifteen meter novice band for signals whose speed suggests that the operators are not novices.

Above all, we are patient; waiting, watching, listening. We are DX'ers.

9
Advanced Pile-Up Technique

There are two especially useful techniques in pile-up situations; the tail-end and the delayed call. Both of these tools are special favorites of the QRP set; however by no means is their usefulness limited to QRP operation. But each of these tools is very different from the other, each having its own place. And each of them, especially the tail-end, can do a great deal more harm than good if used improperly.

...It's Saturday morning, a rainy day in June, but a gentle rain with no local thunderstorm activity. The soft air smells fresh after a long spell of dry weather, and the power line noise that has been a steady S-5 to the West is gone. I'm tuning twenty meters, hoping to find a goody somewhere. There is hope; while there appears to be no long-path propagation to the Middle-East, I hear several strong southern Africans, a very strong VK6 in Perth, and scores of JA signals. The antenna is set on 290 degrees, splitting the bearings between South-East Asia and the long path into Southern Africa and the Indian Ocean.

As I cross 14,030 kHz, I strike paydirt, "R 73 QRZ UP 5 DE V85CM K" It's Brunei! I worked a V85 several years ago, but never could snag anything resembling a QSL, in spite of repeated efforts, so this one is as good as a new one for me. And joy of joys! A sharp operator with an excellent fist, not very strong but Q-5. Now all I need to do is to work this fellow. And even as I turn the linear on, I hear more good news, "R QSL VIA KØHGB 73 QRZ DE V85CM K"

All right! With a QSL manager to boot. Needless to say, I am eager to put this one into the log. I move my transmit frequency up the five kilohertz the V85 specifies, and pause just for a moment before starting my call. Suddenly, it becomes very apparent I am not the only station to regard the V85 as a desirable log entry; a snarling pile-up of loud stations is calling, nearly all at rather high speeds.

I give a quick call none the less, then switch back to the V85's frequency. Yes, he's just coming back, "K9MM 599 K". K9MM is a local; quickly I switch VFO's to listen to my transmit frequency and tune higher, in time to find K9MM transmitting a brief report. But not so brief that I am unable to get set up on his frequency. I switch back to the V85's frequency, just in time to hear his QRZ. I call, a fast 2X call, and listen. Nope, K5VT has him. I jump the receiver back up, to find K5VT. Yes, there he is, not very strong, but I'm on the edge of his antenna. He's just a hair above where I was calling; I move again, setting up slightly higher than his transmitting frequency. And, even as I do so, I hear "DE N4AR", even though K5VT is still transmitting. You'd think N4AR would know better, being one of the real big time DX'ers on any band.

I flip back to the other VFO, just in time to hear the V85 signing clear with K5VT. I'm ready. But wait; he's not calling QRZ, he's calling N4AR! Wow! Are some people ever lucky. The V85 ought to be mad at N4AR for QRMing K5VT's call sign as K5VT was signing clear, even if N4AR was a couple hundred hertz above K5VT. Instead he's working N4AR. There, he turns it to N4AR. OK, quick, switch and find N4AR. Right, there he is, right where I heard him before. OK, set up a hair above him. Gosh, I was almost dead zero with N4AR. Maybe I would have beat him out if he had called at the regular turn. Huh, look at that, K8MFO calling now, and N4AR isn't done signing either. Doesn't anyone around here ever wait their turn?

OK, there, N4AR clears. Quick, switch to the V85's frequency. Yes, there, he's signing clear, saying "73 SK NW K8MFO TNX 599 BK". Darn! Something's funny here. Those fellows are getting through, and doing it without waiting. And that V85 is sharp. He's picking up their one X (one time) calls, even though the fellow that the V85 is already in QSO with is still transmitting, signing clear. Of course, those fellows are transmitting several hundred hertz off the station in QSO, and it is true that the V85 already has the call of the fellow he is in QSO with.

Hey! Maybe that's not such a dumb idea after all. They are not giving the V85 any QRM in copying the station he is working. And they are just dropping in their calls one time, not hurting anybody. What the heck, I'm going to give that a try. Oops, too late. He's finished with K8MFO now. But yes, he's got another one. Yup, it's W9VW. Yeah, that fellow is a wiley old fox. He knows what he's doing, for sure. There must be something to this bit after all.

OK, the V85 is back to 9VW now. Find him. Yup, there he is. OK, set the transmit frequency a bit above him. OK, let's listen. Yes, there, he's starting to sign, I better start calling. Even as I reach for the paddle, though, I hear a fast "DE WB4OSN". Oh oh. Bet I blew that one. I sign my call, and listen back on the V85's frequency. Yup, darn it, I was right, there he is coming back to the WB4. But, I know what's happening now, and I'm ready.

I hop back to WB4OSN's frequency. Yes, I was almost exactly right about the frequency. The V85 seems to be slowly moving up, a couple hundred hertz per QSO. I ease the VFO up, and listen to 4OSN giving the V85 his report. Yes, he's going fast too. There, he's starting to say 73. Now! "DE W9KNI" That's all, nothing more. Switch back to the V85's frequency. Let's see how I did.

"R 73 SK NW W9KNI DE V85CM GM ES TNX 5NN BK". All right!

"R V85CM 5NN 5NN 73 DE W9KNI SK"

Wow! Was that a fast one. But there, "R W9KNI TNX 73

THE COMPLETE DX'ER 79

ES QSL SK NW K6NA 5NN..." Yup, he's in my log, and I'm in his. So that's tail-ending, huh? That's all right. I'm going to have to try that some more. Let's go look for something else to work.

I'm like a kid with a new toy! I move up the band. There's a VK3 in QSO with a W4; I don't need a VK, but let's try a tail end call on him and see what happens, soon as he signs with the 4. A little practice never hurt anyone. Yes, there, he's signing, and turning it to the W4 for his final. OK, there's the W4, I set the transmit VFO up a hair above him. Right, there, he's signing clear; I drop in "DE W9KNI" and sit back.

Hah! It worked again. Where have I been all my life? This tail-end business is the greatest. I have a brief QSO with the VK, and I'm off looking for new worlds to conquer. Yes, there's a nice one; a P29, with several stations hot on his tail, but he's coming back to one of them, yes, a WB8. Oh-kay. I don't need a P29; I have one on the wall, but no harm showing the troops how it's done, heh heh. This ought to be like shooting fish in a barrel. The W8 is not very strong on backscatter, but plenty strong enough for me to set up my transmit frequency just above him.

The P29 is not going fast, and is exchanging complete QSO information with the WB8, but soon they are signing with each other. I slow my keyer down a bit, slightly faster than the speed of the P29, just like those fellows were doing with the V85 earlier, and just as the WB8 starts his call, I neatly drop in my call, "DE W9KNI" and sit back, picking up my pencil for the log info.

Hah! "WB8EUN DE P29JK OK DARYL TNX QSO GNG QRT NW W9KNI QRMED U ES I WILL NOT QSO W9KNI AND I QRT INSTEAD 73 ES GOOD DX WB8EUN DE P29JK SK QRT CL"

I feel like a 3-500Z being used for teletype at a KW. My face is red, and hot. How do you crawl under a rock on twenty meters, with a pile-up of frustrated DX'ers writing down your call in the margin of their logs. You know, maybe that rain outside has let up enough for me to go out and pick up the yard a bit. Somehow, I don't feel like operating much today.

Or, as the old Bard said, "Live by the tail-end; die by the tail end." And sure enough; I died. Sure am glad I don't need a P29, though...

THE COMPLETE DX'ER   81

Like many aspects of DX'ing, the tail-end call is a two-edged sword. It can be a marvelous tool for knifing through a pile-up, or a good way to get knifed. As is always the case, the final verdict rests with the individual DX station, not in some rule book. Some DX stations are very happy to pick up tail-end calls. Others ignore them. And a few get very upset with them.

Unfortunately, there is no simple way to determine whether or not to try the tail-end. The safest thing to do is to let some one else try the technique first. But, sometimes, the sacrificial lamb fails to appear in a timely manner. And, as you listen, the band is slowly going out and the pile-up is increasing rapidly. What to do?

There are several guidelines. The first is that if you think that you can get the QSO without using the tail-end, then don't try using it, unless you know that the DX station is accepting tail-end calls. Why risk angering the DX station?

If the DX station is a good operator, and sending at a high speed, it is usually fairly safe to try a tail-end, IF you time it properly, and IF you spot your frequency accurately enough to be sure that you are not causing QRM. If you can't spot your transmit frequency that closely, don't even THINK about tail-ending unless you are seeking a reputation of other than honor roll stature.

Tail-ending is often effective with Soviet stations; it seems to be a technique widely used by them for their intramural QSO's.

If you still are unsure in a given situation, there are several techniques worth trying that can limit potential damage. One is to go QRP on tail-end calls, QRP being in this case at least 10 dB down from the fellows getting QSO's with the DX station. Also, be sure that your transmit frequency is a bit further off the station in QSO than you otherwise would, say 300 hertz instead of 150 to 200.

There is one other possible benefit of tail-ending. Often, DX operators will not accept tail-end calls because they do not want to encourage the use of them, but they are not upset by a properly executed attempt. This situation gives you a chance to "show" your call and frequency before the pile-up starts. If you suspect this to be the situation, try careful calls 300 hertz off the QSO'ing station's frequency, and do not move frequency during the pile-up call, but call as you normally would in a pile-up. There is a good chance the DX station will pick your call up, having heard your tail-end. However, if this trick fails to bear fruit after three shots at it, quit! You may be running a real risk of getting on the DX operator's black list.

Tail-ending, then, can be a valuable weapon in your kit bag. But it should be used conservatively. Unless you know for a fact that the DX station is accepting tail-end calls, attempts should be made very sparingly. The QRPp operator can use it at will; his five watts is unlikely to anger the DX station, and he has little to lose. All other operators beware!

There is one other special technique that can be of great value in pile-ups, and unlike the tail-end call it is a technique that usually will not get you in trouble. This technique is known as the delayed call. It is not a trick for every one; proper use requires steel nerves, and it runs counter to most people's idea of how to operate in a pile-up. That, of course, is why it can be successful...

Band conditions are pretty poor, but you certainly aren't getting a complaint from me, even if it is a vacation day in mid-week. Because, you see, about twenty minutes ago I caught ZD9HL calling CQ, nice and slow, big and fat. I was all over him like a ton of bricks, and soon writing the QSO details into the log. A new country! How sweet it is.

Now, with little hope of anything else good showing up, I am sitting on the frequency, listening to the rapidly growing pile-up. The ZD9 appears to be a new operator, working slowly and carefully, a bit tentative. He does not seem to be moving his receiver much, either.

It's amazing how quickly a pile-up can form on a rare one, even in mid-week on a crummy band. As I listen, I hear more and more stations entering the fray. The ZD9 is certainly aware of this; now he has dispensed with much of the QSO information, and is more or less down to reports. And with each QSO the pile-up mounts.

As I sit listening, I start watching carefully, trying to see what techniques those stations who get through are using. In fact, I pretend that I am the ZD9. Of course, signal strengths are not the same for me as they would be for the ZD9, but it is still an interesting exercise.

It quickly becomes apparent that the primary interest of

THE COMPLETE DX'ER 83

the ZD9 is to find stations on that he can copy one way or another. It appears that he must be using an old model transceiver without RIT, since he seems unable to move his receive frequency. Everyone that he works is within a few hundred hertz of my original transmit frequency.

However, many of the stations calling him seem unable to place their transmit frequency inside that relatively narrow slot, and their calls are in vain. And within the narrow bandpass that the ZD9 is listening to, it seems that he is moving his ear around a bit. The main criterion for getting a QSO seems to be to transmit on the ZD9's listening frequency with no one else zeroed on you.

I make note of the calls that seem to be on the right track. I notice ole' Louie, K5LM, calling in there. Louie is an old friend, and noted for being a superb tactician in the pile-up, so naturally I listen with special interest, waiting to see how he proposes to solve this puzzle.

The situation appears to be slowly deteriorating, even though the ZD9 is struggling to do the best he can. The QRM is rapidly building, and the ZD9 is taking longer and longer to come back to each new QSO. Obviously, his equipment, and perhaps lack of experience, are not helping the situation.

I listen to the QSO's, glad to have the fellow in the log. There, he's signing clear with WB4CSK. Listen to that pile-up. Yes, there's Louie, right in the middle of the slot the ZD9 is taking calls out of. OK, the ZD9 is back, to another 4, an N4... Right, he's done. There they go again. Louie has moved down perhaps 200 hertz. Nope, that didn't work for him either. The ZD9 is back to an 8, a KB8... OK, they signed. There goes the heap again. Where's Louie? I'm not hearing him. Boy, listen to those fellows all nearly zero beat QRM'ing each other in the ZD9's listening slot. But I don't hear Louie.

There, the pile-up is thinning out, and the ZD9 still isn't back to anyone. Hey, there's Louie, and on a different frequency, too; "DE K5LM K5LM AR". Gosh, he had a clear frequency for a moment, right after all those other stations signed... And there! The ZD9 is coming back to Louie! Way to go, Louie! All right!...

That was really neat. When Louie figured out the scene, he quit calling while everyone else kept on, and laid back in the bullrushes, his hand on the dial. As soon as the pile-up started to let up, he found a frequency that cleared for a moment, quickly dropped his transmit VFO on it, and got that clear call. That is the delayed call technique. And it worked.

The technique can be a very useful one indeed where the DX station is slow in coming back. Not every DX station has super crystal filters cascaded, and waiting for a frequency to clear is often the only way that he can copy a call. This means, in turn, that the astute DX'er can lay back, waiting for that frequency to clear, and grab it for his call.

Obviously, to make this technique work, the DX'er needs flawless timing, the ability to quickly and accurately zero, and the patience to wait for his opening. This means not calling the DX with the rest of the pile-up, but rather holding back, waiting for the opening.

The delayed call is an equally important technique for the SSB DX'er, especially where the DX station is working stations only on frequency. A few moments pause until all stations on frequency have made their calls will frequently offer a brief moment where the QRM has died down, but before the DX has been able to copy a call, giving you a chance for a quick call.

The reader will also have noticed another useful technique here: listening as if you were the DX. Putting yourself under the DX station's headphones is a good way to see what techniques are effective, both for that specific station, and for working DX in general. The good DX'er is always trying to learn.

10
CQ DX CQ DX CQ DX DE...

One of the most vilified techniques of DX'ing is the CQ DX call, yet, this situation is very unfair. True, for the QRP operator, calling CQ is most often an exercise in futility. But for the operator with an anemic country count and a decent signal, it can be an effective way of fattening one's country count in a hurry. Also, there are occasions where it is an excellent technique for snagging a particular station, but this aspect will be covered later in the book.

There are a number of reasons why a CQ DX call can be an effective tool. For example, some rare DX stations do enjoy QSO's that are more meaningful than the usual hit and run types. If they respond to a station's CQ DX call, they hope for a proper QSO that ignores the rareness of the DX station.

And, a QSO of this type, with perhaps some interesting exchanges, is far more likely to generate the needed QSL later. It can even lead to an interesting friendship.

Another reason for a CQ DX call can be to help stimulate a little action on a path that is open but not in use. If you have a directive antenna with gain, you should by now be well aware of the many secondary paths that open up, particularly on twenty and fifteen meters. Some of these paths can be to rather exotic areas of the world, where almost any station qualifies as rare DX, encouraging such stations to lie low if they are not in the mood to deal with a snarling pile-up.

Some rare DX stations have a situation where they are close to a country with a large and active amateur population, so that any time they try a CQ of their own they are sure to be buried in calls from the nearby amateurs. An excellent example of this would be a station in the Philippines, or in Hong Kong, in easy skip distance of the huge Japanese DX community. If the VS6 or the DU does not want an instant pile-up of JA stations, he avoids CQ calls, but if he is tuning the bands looking for a QSO and hears a nice strong W signal calling CQ DX, there is an excellent chance the W station is going to get a rather nice log entry out of his CQ.

Another situation where CQ calls can raise some excellent DX is during an international DX contest, especially the CQ WW contest, the All Asian contest, and similar contests. As will be discussed later, however, there are so many good opportunities to snag new ones in such a contest that the DX'er is usually better off looking for the new ones rather than waiting for them to come to him. Still, in such contests there are invariably rare stations operating that can only be raised on a CQ, so even then some CQ calls can be indicated.

...It's a late August Saturday afternoon, and for a change the chores are well in hand. The Cubs are in New York riding a nine game winning streak, but it's a night game. (After all, there ARE other things besides DX). A look over twenty meters shows it to be in decent condition, with a good contingent of European signals coming through with nice signal strengths. But I only need Albania, Mt. Athos, and Monaco from Europe. I suppose there is a slim chance of a 3A showing, but I am not holding my breath. And my chances of tripping across a genuine ZA station or a Mount Athos operation are minimal.

THE COMPLETE DX'ER

I swing the antenna around to the north-west, across Japan and into S. E. Asia. Careful listening unearths a couple JA stations with S-7 signals. What the heck, I decide to try a CQ. I find a clear frequency, and confirm it with a "QRL?" which goes unchallenged. I set the keyer at a nice comfortable speed, not too fast, and transmit, "CQ DX CQ DX CQ DX CQ DX DE W9KNI W9KNI CQ DX CQ DX CQ DX CQ DX DE W9KNI W9KNI W9KNI CQ DX CQ DX CQ DX CQ DX DE W9KNI W9KNI W9KNI DX AR K"

I open the receiver. There, a couple stations are calling me. OK, one signs his call first, OK, it's a JA. Let's see who the other one is, maybe I'll get luckier there. Nope, it's a JH3. Oh well, I'll take the JA. I respond to the JA, and we have a nice though brief QSO. Taka's English seems a bit limited, but I can't complain. It's far superior to my Japanese. Soon, after promises of QSL's, we sign, and I go fishing again, with a "QRZ DX QRZ DX DE W9KNI W9KNI AR".

Again, two signals calling me, with possibly a third at the edge of my pass band. I move the dial a hair; yes, another station calling me. I start copying calls; the JH3 that called me last time, a JI1, and the signal off to the side a KA7. I pick up the JH3, after all, he was willing to wait. I have another QSO just like the previous one. Again, when finished, I call, "QRZ DX QRZ DX DE W9KNI W9KNI".

The path must be pretty decent, especially considering that it's a secondary path; I hear four or five stations calling me. Again, I try to sort out the calls, looking for the cream that I'm hoping will rise to the top. Hello! There's a different call. Yes, it's an HL9, Korea. There's one that's not all that common, even though I have one. Yes, that's more like it. I'm always willing to talk to the JA's, but something more exotic adds a little spice.

The HL9 turns out to be an American serviceman stationed in Korea, and from the Midwest. He seems glad to get the news of the weather, and we discuss the Cubs, wondering if their lead in the National League East over the Mets will hold up. After about fifteen minutes of pleasant exchanges, he has to QRT for duty.

It appears that the band must be changing now. There is only one signal calling me, and I wonder if the path to Japan has faded out. Secondary paths can be very selective that way. There, the station calling me is signing his call; all right!

9M6SP! East Malaysia, and a gem! A new one for me. Huh, I didn't know anyone was anybody active from there.

Even though I'm in a mild state of shock, my trusty Bencher paddle helps me through my transmission. He responds, having copied solidly, and, after the preliminaries, we begin to ragchew. His name is John, and he is in a small village near a town called Sibu. He tells me of the weather and his rig. He is using a home-made half-wave vertical for an antenna, with open wire feeders and a bamboo mast. He tells me about his gear; old war surplus stuff.

It seems that he is an expatriate Englishman in Malaysia working as a teacher. He tells me about the school he teaches in, a secondary school, and he teaches mathematics and physics. He has been in East Malasia for fifteen years, having gone there as a soldier, and deciding to return after his enlistment was over. He is married, and has two children.

He was very glad to hear my CQ, because he likes DX QSO's, but with the old BC-348 receiver there is no way that he can handle pile-up situations, so he only QSO's W stations by answering CQ calls. His transmitter is war surplus as well, a much modified BC-458 using a pair of 1625's. His VFO is interesting; a BC-454 receiver modified into a VFO mixed with a crystal controlled oscillator. And the antenna tuner is fabricated from another war surplus rig too!

I ask about the paddle and keyer he is using. All homebrew, too, the keyer from parts purchased mail-order; the paddle from a pair of J-38 (you guessed it - war surplus) hand-keys back to back making an iambic paddle.

Needless to say, I am impressed. I glance over my gear. Hardly a comparison. Not to mention the very modern array perched atop the sixty foot tower, and the 3-500Z afterburner hooked to the coax. But there he is, with a good clean signal, a pleasure to copy.

I ask if he needs any spare tubes, having some old metal tubes in the junk box that may well serve him better than me. He is very excited at that news. It seems that he has no spares for several tubes, a 6AG7 and a 6F6. I laugh; I have plenty of both. I promise to have a care package on its way within the week.

His signals begin to fade, slowly but surely, as he finishes a long transmission. I tell him that I shall be sending him the tubes along with my QSL, and propose a schedule for the following week-end. I pass along 73 and the usual closings, and turn it back. He is definitely weaker now, but I am still able to copy solid. He agrees to meet me a week hence, and thanks me again for the tubes. Also, he promises to QSL via air mail on Monday, so I can be looking for a fast one there.

He signs clear as his signal sinks into the mud. About twenty American stations start frantically calling him, but as I expected, he comes back to no one. I hear, very weakly, JA stations calling him as well, but after what he said earlier, I am fairly confident he is gone. And he is.

I sit back, thinking. Wow. A new country on a CQ call, a rare one. But even more, a QSO that is truly one of the more memorable of my amateur career. An exotic spot on the Isle of Borneo, with a fellow from a common heritage, but leading a far different life. And the gear he was using! It all makes me proud be a ham and a DX'er, letting me work interesting people like him.

I go up the basement stairs, jealous of John in Borneo. I imagine that his life is more interesting, and rewarding. And I dream of tropic palms and bamboo thickets. But we are what we are, and what will be will be. And if I can't go to exotic places, and do interesting things, at least I can talk to people who do. And keep chasing new countries...

The QSO described above is a kind that will never come of winning a pile-up. And it is a sort of QSO that, unfortunately many DX'ers have never experienced. Yet, such QSO's are far more common than most amateurs will believe. There are significant numbers of stations operational in areas considered rare, like John on Borneo, operators that lie low because they lack the modern gear capable of coping with pile-ups.

There are a number of other ways to smoke some of these DX stations out, and this subject will be further discussed later, but suffice it to say that a CQ DX call is one good way to do it. If you are running 150 watts to a mediocre dipole, CQ DX calls will serve to do little more than enhance the profits of the electric company. But, if you have a good directive antenna such as a quad or a yagi, at fifty feet or better, your chances are pretty good, even running bare-foot, and of course are enhanced if you are running the legal power limit.

CQ DX calls can certainly be aimed at the major path of

the band, but you are likely to be far better rewarded trying the secondary paths of twenty or fifteen meters. For one thing, to the DX station at the other end, you may well be the only W signal on the band, since all the other W stations usually have their antennas on the primary path opening. Remember, it is new countries you seek, not reports from countries you have worked a hundred times.

Also, when you are calling CQ DX, you are sure to raise a lot of responses from stations you certainly don't need, such as JA stations on the path to SE Asia. Fairness says that you must work these stations as well if you don't get a call from a rarer one; but indeed the author's experience has been that the rare ones come back more often on the QRZ DX calls at the end of a QSO rather than to the initial CQ. Probably this is partly because the rare DX doesn't wish to transmit until he is sure that you are listening in his direction, as he seeks to avoid raising a pile-up of stations more local to him.

To sum it up, if conditions seem to be favorable, and you are finding nothing new on the bands after a careful tune, give the CQ DX technique a try. And if it doesn't unearth a rare one at first go, don't give up; it is a technique that requires patience. And it is like any other technique of DX'ing; the more time you put into it, the better rewarded you will be.

11
CONTEST!

Contests offer the serious DX'er an unparalleled opportunity to grab a lot of countries, often including a number of rare ones, in a very short time. Additionally, all the aspects of successful DX'ing are crammed into one or two week-end days. There are towering pile-ups to be fought, often over stations that on any other day would hardly raise an eyebrow. Rare DX lurks between the big guns and their mindless CQ DX machines. Contests are indeed a microcosm of DX'ing, and offer the committed country hunter a rare chance to practice his techniques, sharpen his skills, and snag some new ones.

However, the DX'er must first decide whether he is a DX'er or a contester. There is a very significant difference. The DX'er uses a contest strictly as a vehicle to grab new countries, improve chances of getting QSL's not already received, and sharpen his skills, both in tuning for the rare ones and slugging it out in the pile-ups. He is NOT tempted by meaningless QSO's just to fill lines in the log sheet.

This does not mean, of course, that there is anything wrong with contesters; indeed it is thanks to them that DX'ers get such wonderful DX'ing opportunities. And those that love contests are equally brother amateurs. But, the goals of the contester and of the DX'er are much different in contests. The contester generally works every station he can find that he is allowed to work, and additionally becomes involved in pile-ups over stations that, outside of the contest format, would have no particular value.

We presume that your primary interest in the contest is that of the confirmed country hunter. There are several contests that are especially important to the DX'er. Generally, the far and away best for the DX'er is the CQ WW contest, with the Phone weekend in late October, and the CW weekend in late November, generally just after the American Thanksgiving Day. Other important contests include the ARRL DX contest in early spring, the summer All Asia contest run by the Japanese, and the Soviet CQ-M contest early in May.

The early hours of a major contest are generally chaotic, with the "big-gun" multi-multi contest stations intent on throwing their weight around and monopolizing frequencies. Techniques practiced by some of these worthy brother amateurs include deliberate QRM from secondary antennas turned away from the DX paths, incessant CQ calls to protect "their" frequency, intentional key clicks to keep others from getting too close, power levels that honor their license regulations only in the breech, and a general level of behavior that disgusts other contesters. Certainly, this is not true of all multi-multi's. But, tragically, the stations that are most effective in these techniques are those that usually win.

The DX'er in the early hours of the contest generally has his best luck working on the secondary paths, staying away from the major openings. If fifteen meters is wide open to Japan, the astute DX'er will be checking it out for the Antarctic islands, and watching the trans-polar path on twenty. In other words, avoid the mob scene, and tune between the pile-ups.

Generally, on Saturday morning the bands tend to get

shaken out. The multi-multi's have staked out "their" frequencies, and woe to those who trod near them. Usually, a bit higher up the band, things begin to open up a bit...

I slept very well last night. Fifteen had been pretty good last night until about 0200Z, when it pretty well closed down. I had been hoping for a bit of an opening into the South Pacific, but we lost propagation before activity really hit its stride. There had been loads of Japanese stations earlier, but I had pretty well ignored them. I already have Japan confirmed.

None of the Antarctic islands showed. I had checked twenty meters over, especially keeping a watch for any central African stations, but the search went for naught. When twenty went flat, I packed it in and headed for the sack, deciding to let the contest gang fight over forty and eighty. It had been a hard week, and I wanted to start Saturday morning fresh.

Now it's a few minutes after 1300Z. I am seated at the rig, hot coffee near at hand, starting to tune twenty. If conditions are any good at all, most of the contest gang will have switched to ten and fifteen meters, looking for the multipliers from Europe, leaving only the multi-multi robots tearing up the bottom of twenty. But in the meantime, there is still more frequency than the multi-multi CQ machines can fill, and hope for long path Middle-East DX, plus South-East Asia. I park the antenna South-West, so I can check the long path first, and start tuning up the band.

I quickly move out of the bottom twenty kilohertz of the band; it is a zoo, and there's virtually no hope of finding any weak signal DX. The loud signals obliterate my AVC system, and the key clicks of the CQ machines make weak signal copy impossible. But above around 14,020 kilohertz sanity begins to become apparent again, and I am able to copy weaker signals. I hear East Europeans, an LZ and a YO, along with regular UF6CR banging away. Their signals are S-5 to S-6, an indication that there is at least some propagation via the long path.

I hardly go further before I strike my first bit of luck; a small pile-up over someone. There, who ever it is is transmitting, "5NN21 5NN21 K". Hmm. Zone 21 is a Mid-East zone, which guarantees interest. Someone comes back, "R 5NN05 5NN05 73 BK". The DX station responds, "R TU QRZ DE HZ1HZ K". Hah! An HZ1 in Saudi Arabia. And, a good QSL'er as well, apparently, because I have seen his QSL on the walls of several of the local DX'ers shacks. Also, it would be a new one for me!

Needless to say, I quickly set up on him and drop in my call. I listen; nope, some one else got him. As I listen, I hear W9VNE drop in a tail-end call. Good! If that works, I'll try it to.

And it does. "73 NW W9VNE 5NN21 K". Even as I hear the HZ sending VNE's call, I am setting my VFO a hair above VNE's transmitting frequency. There, VNE is back. As soon as he has given his zone number to the HZ, I drop in my call, just once, "W9KNI", and wait. And Bang! I get him! "W9KNI R 5NN21 BK" I'm right back at him, "R TU 5NN04 73 SK". And there isn't anything slow about him either, "R TU QRZ HZ1HZ K". I grin from ear to ear, check to see that I have the log data correct, especially time and date, and tune on. They should all be so easy.

Needless to say, my continued tuning is done with particular fervor; but in due course it appears that there are no further Mid-East stations worthy of mention to chase. I swing the antenna further south to 180 degrees, in hopes that the transpolar long path might be in. After all, I can always use the odd VU7 or A51.

The path does prove to be open; almost immediately I come across a VU2 spraying out reports in an expert fashion. And, in terms of mileage and path difficulty, VU is always a good DX catch. But I have it confirmed, so I tune on. After nailing the HZ1, I am hot for more new ones to feed the log book. Still, the path has its interests. I hear UH8DK calling CQ. Although I have five UH8's logged, I have not yet gotten any QSL's from Box 88. I check my short list, and see that UH8DK is not one of the ones that I have worked, so I zero him and call. I fail on the first call, losing to a GW3 station I can only detect in the background, but when the UH8 goes QRZ, I get him. Soon, he's in the log, and I am moving along again, still checking the polar long path.

The path seems in good shape, as I find a 4S7, several more VU's and the odd Soviet station, but I come across nothing else that I need. I turn the antenna into the North-West, to try my hand for South-East Asia, and keep turning the dial. In a contest, most DX'ers tune faster than they might in normal operation, because the normal events of weeks of activity are boiled down, as it were, into 48 hours. In non-contest circumstances a careful DX'er might watch a good open long path for an hour or more, while in a contest he tries to check that path in a few minutes, going back to it regularly to check. This is necessary partly because the normal clues a good DX'er watches, such as pile-ups over rare ones, tend to be obscured by contest conditions.

I move my way up the band again, using 14,020 as my starting point, rather than 14,000 as I would in normal times. Almost immediately, I come across a strong JA station, but with that peculiar flutter that occurs when only the stations with high large directive antennas are heard, while the stations running ground planes and normal dipoles are not to be found at all. But that's not surprising for this time of year; that path doesn't usually open as far north as Japan in the morning.

I find a YB5 with nice strength passing out reports; that's a good sign for the South-East Asians. And seven kilohertz above him a 9M2 is dealing to a large following. I glance at the clock; it's nearly 1500 Zulu; better keep moving; I want to have a look at fifteen pretty soon for some Africans. Also, for us in the American Midwest, any chance of working SE Asia over the W6's and W7's is about gone; if we are to make hay on that path, we have to do it a bit earlier, before the West Coast gains a daylight propagation terminal. I finish tuning over twenty without further incident of note, turn the antenna East, and head for fifteen meters.

Fifteen turns out to be a friendly madhouse, a hundred and fifty kilohertz of sheer bedlam. The band is open very nicely, and the bedlam suggests that the MUF never quite made it up to ten meters. But no matter, fifteen is a joy, loaded with loud European signals. The only trouble is, I don't really expect any of them to be from Albania. But, in any case, it's not Europeans I am after.

I start tuning the band, looking primarily for good Africans. The antenna is due east, which will give me good signals from Africa, but unfortunately the Europeans remain fairly strong. I try moving the antenna further to the South, while I listen to an ON4 calling CQ. His signal drops rather sharply as I move the antenna, and when I have the rotor at 110 degrees he is quite weak.

I quickly check my Second Op. TN8, for example, is 83 degrees from the mid-west. My 110 degree bearing is 27 degrees off from that, still plenty close. I probably would lose only a single dB in receiving strength on a TN8, but I drop the European signals twenty dB or more. And if I should get lucky enough to find a TN8, I can always quickly get the antenna back to the 83 degree bearing. And in the mean time a TN8 would be a lot easier to find with the European signals attenuated. I am making the presumption that a TN8 would likely be at least S-7, an implication that would be suspect on twenty or forty meters. But this is a contest, on fifteen meters, and different rules apply for efficient DX hunting.

I start tuning again, hoping to find one of the "Terrible T's". (TL, TN, TR, TT, TU, TY and TZ, all generally rare.) Nor would I spurn the advances of a 5U7 or a 3X. My first quick pass turns up nothing. I am hoping to do some good for an African while fifteen is still hot to Europe, and the masses of contesters are still concentrating on them. Later in the afternoon, after the Europeans fade out, pile-ups on good Africans can assume monumental proportions. I would prefer to work my new ones now, ahead of the madding hordes.

The second time through the band, things get more interesting. As I tune across a loud W station, I suddenly realize that the call he just sent is TL8GI, one that I need badly. (Just like all the other countries I still need!) I listen; yes, there he is, S-8, too, "R 5NN36 K" and, with the other station's acknowledgement, he's gone. The other station calls CQ, telling me that the TL8 answered his CQ. I hang my VFO 300 hertz above the QSO frequency, and try calling the TL8, but to no avail. I call again; same luck. I'm in a panic.

I move up, looking, while I swing the antenna back north about 30 degrees. Almost immediately, I find him, about six kilohertz higher. I don't hear his call, but I hear him give a report, the "5NN36 5NN36", and I'm pretty sure it must be him. If nothing else, zone 36 is rather rare. When the other station clears, another station starts calling the TL8, confirming my suspicions, but the original station is calling QRZ DX, and is not about to surrender his frequency.

I quickly look a bit higher, and find a relatively clear frequency about 4 kilohertz higher. I figure the TL8 keeps moving higher after each QSO, so I try a fast CQ call. Someone comes back to me, and for a moment my heart jumps; the signal strength and speed seem correct, but it turns out to be a VP9. I give him a quick report, then call QRZ DX. Nothing. I move the receiver again, looking.

There, several stations calling him, about five kilohertz above my last transmit frequency. I already know he won't be sticking around; clearly this fellow does not care at all for pile-ups. Probably that's why the place is so rare. I quickly move up past them, and find another frequency that is not too busy six or seven kilohertz higher. I dump out another CQ call, and get two stations coming back.

I will never know who one station that called was, because the other is the TL8! The contact goes like lightning; "TL8GI R TNX 5NN04 5NN04 PSE UR QSL K". He responds, "R 5NN36 5NN36 QSL K5OA K5OA 73 SK". And that's all there is. But it's enough. Wow, another new country! And with a QSL manager, too! Contests sure are fun!...

And that's how contests can go. The example with the TL8 is in fact representative of a fairly common situation in contests - rare DX stations that want to avoid pile-ups, but do want to get in on the action. Such stations are most often found in the CQ WW Contest, but it can happen in any contest, and the DX'er must be ready. Sometimes the DX station will respond to a quick call perhaps 300 hertz above the station he just worked, and, if possible, the DX'er should try this. If it is unsuccessful, as it was in the situation above, a planted CQ, as above, is the only hope, and, in fact, will work a high percentage of the time if done correctly.

Even for the DX'er with 300 countries and little hope of snagging a new one, DX contests offer exceptional opportunities to practice pile-up tactics, and to keep a sharp edge. And, if pile-up practice is the name of the game, the DX'er should consider competing barefoot, rather than with his big linear. It gives a bit of a handicap that is desirable for practice, not to mention making the neighbors grateful!

Contests are a great place to sharpen techniques such as the tail-end call, the delayed call, precision spotted calls etc., all for stations that are not needed for new ones. But always remember too, if you are in a pile-up over a station that you don't need, you can't expect to work one you do need who may be just up the band.

12
Morning Call

I pull the chair up closer to the table, and snug the earphones tightly over my ears. I got lucky today; a fine frosty Wednesday in mid-December and me with an appointment ten minutes from home at nine AM. I've already bathed and dressed for the occasion, needing only to put my tie on for a last minute dash out the door, and now I can enjoy almost an hour of DX'ing.

On the two meter DX net last night, Bob, W9NZM, had referenced a fine long path opening into the Middle-East, with the ever elusive YI1BGD and an apparently legit 4W both available on twenty meter SSB, and both relatively workable. I was surprised not to find my pillow wet from drooling in my dreams. Both of those countries are on my all-time needed list.

I find a clear frequency around 14,200 and tune the rig up for sideband. I check the mike gain, adjusting it with a very brief test transmission, point the antenna southwest, and set the transceiver at 14,150. I pick up the two meter microphone and give Bob a call. "W9NZM, here's W9KNI."

"Good morning Robert. Here's W9NZM."

"Hello Bob. Say, I'm going to be able to do a little hunting here for about an hour. Anything good on?"

"HZ1AB is coming through on fourteen one eighty-eight like a ton of bricks, but that's it so far. The band is certainly open."

"Good. Say, what can you tell me about that Four Whiskey?"

"His call was Four Whiskey Zero Kilo Portugal. He spoke decent English and seemed like he knew what he was doing. He was transmitting on about two-ten."

"OK, thanks. I'm looking for the YI also. Let me know if you hear anything."

"Roger, Bob. Good hunting."

"Thanks." I put the two meter microphone down and begin a slow tune up the band.

Almost immediately I come across a station speaking English with a crisp accent that is decidedly non-American, with a good 5X7 signal. "...name here is Clive, Canada, Lima, India, Victoria, England, Clive, and the QTH is sunny Perth, on the western shores of Australia. Back to you, Old Man. Whiskey Alpha Four Germany Tango Foxtrot, this is Victor Kilo Six Mexico Lima, over." I hardly need a VK, but it is nice to hear such good signals from Western Australia, indicating good band conditions, as Bob suggested. I keep tuning.

I hear a Five calling CQ DX, speaking slowly and clearly into his mike. He is only S4 with definite backscatter, suggesting skip is long, and trouble with stateside interference will be minimal. I hope.

I listen to the Five finish his call, wanting to see what he dredges up. He signs, and two stations begin calling him. One is a good catch; a JY9. I strain to hear the other call, and manage to pull it out; it's an FR5, not as strong as the JY. That would be a nice catch too, but not one I need. The Five starts responding to the JY9, and I continue tuning higher up the band. As I do I think about the nuances of the morning twenty meter long path opening. The good news is that almost every station coming through on that path is likely to be fairly rare DX. The bad news is that the competition can be really fierce.

But, as with any DX band and opening, many operators simply tune for the pileup, never giving a look at the weaker signals leaking through on the band. Which gives the careful listener the opportunity to unearth goodies ahead of the pileup pounders, and before the would be list-takers show up to try to wrest away the DX station...

I come across a loud signal, still almost two kilohertz away, and automatically start tuning him in. It proves to be a W4, who is obviously talking to an old acquaintance. He lets up for a moment, and I think I hear a weak signal back below him, a frequency I had crossed in tuning in the Four. I tune back to the other signal, but the W4 keeps talking, obliterating the weaker signal. I try turning on the narrow filter, but to no avail. I sharpen up the audio filter, and manage to figure out the station is calling CQ, and that there is a "W" in his call, as I catch a definite "Whiskey" in the phonetics. Of course, I fantasize that the "Whiskey" is a part of a "Four Whiskey Zero..." Then, the station quits his call, just as the W4 turns it back to the W2, unfortunately having a ragchew right in the middle of the DX portion of an open band.

I listen carefully, trying to see if anyone is calling the mystery station, but I hear nothing. Then, he begins his call again. "CQ..CQ..CQ North America, from 7Q7LW, Seven Quebec..." At that instant, the Four starts transmitting again, burying the 7Q7 as before. But a 7Q7! A rare one - and a potential new one for me, if I can nail him. As I swing the antenna a little more to the west, I listen carefully, trying to piece together bits of his signal in between the waves of RF emanating from the ragchewer, trying to hear when he ends his call. There, he signs... Hope no one else dug him out. I pause briefly; no one else seems to be calling him. I start my call, speaking very slowly, "Lima Whiskey, the Seven Lima Whiskey station, here's Whiskey Nine Kilowatt Norway Italy, Whiskey Nine Kilowatt Norway Italy, over." I don't want to attract attention to him until I'm sure I have him, so, even though I know his full call, I use only the suffix. If I'm fortunate enough to get the QSO, I'll give his full call.

I listen. For a brief moment the frequency is clear, and I hear my call coming back to me! All right! He is giving me a report, which I miss, then a momentary lull in the QRM lets me copy his name as Les. Then the QRM washes over him again and he is obliterated. A moment later, the QRM drops, just in time for me to copy "Whiskey Nine Kilo November Italy From Seven Quebec Seven Lima Whiskey, over."

I return, purposely speaking slowly. "Seven Que Seven Lima Whiskey, here's Whiskey Nine Kilo Norway Italy. OK, Les, thanks for the QSO. Copy is very difficult, due to stateside Que R Mexico. Copy would be much better down one and half kilohertz, Les. May I call you there? Over."

I listen. The frequency is a mess again, but I manage to pick out a "Roger," and I move down. I listen at the new frequency briefly. There is no one on the frequency, with only the usual bits of splatter. I call the 7Q7 again, then, holding my breath, listen. There he is, virtually in the clear, calling me...

Grinning widely, I come back to him, and the QSO turns into a brief but pleasant ragchew. But after ten minutes, we part; he recognizes that the band is open, and wants to work other stations, and I am only too happy to accommodate. After the promise of an exchange of QSL's we sign. I finish my final transmission, then listen with a broad smile as the roof falls in on the frequency, with dozens of stations calling him. I smile. That Four with his rag chew in the middle of the DX band is in for a bit of trouble copying, no doubt. Tough. He belongs up high in the band. Not that he'll ever get the hint. Anyhow, I turn the antenna back into the Southwest again and begin tuning up the band again. Maybe I'll still find the YI...

The above episode points up a very important rule for the DX'er, true for the CW operator, but even more so for the SSB DX'er. ALWAYS strive to copy all weaker signals you tune across. On CW, the natural tendency when coming across a strong station is to tune that signal so that the note is comfortable, even though we might pass a weaker signal in the process. On SSB, that tendency is even greater.

Once we hear a strong station we all share the natural habit of tuning the receiver so that the garbled SSB of that strong signal becomes intelligible. In doing so, we will ignore weaker signals the receiver passes over, signals that may well represent rare DX. It is a habit pattern that makes perfect sense in the normal world, where the person closer to you is more likely speaking to you than a person further away, which is why we have that instinctive pattern. But we are not dealing with the normal world. We are chasing DX.

Once the operator recognizes this natural tendency, it is

not a difficult pattern to break, especially with practice. And indeed, understanding this pattern is one of the major keys to successful DX'ing. Instead of allowing our attention to be "pulled" by stronger stations, we must learn to mentally filter the stronger stations out, not allowing them to distract us, while we try to derive intelligence from the weaker signal, be it on CW or SSB.

The natural tendency to devote our facilities to the stronger signal is all the more aided by putting the AGC or AVC on the "Slow" setting, which further helps to lock out the weak station when an adjacent strong station begins transmitting. Whenever you are tuning for DX, no matter on CW or SSB, ALWAYS use the AGC or AVC in the "fast" position.

Please. Go back and read the above four paragraphs again. And again. To understand them fully is a major key to successful DX'ing. Make a note on the back of a QSL and put it on your shack table. Listen for the weak ones, not the strong ones!

I glance at the clock. I still have the better part of an hour before I must leave for work. I keep on tuning. A few kilohertz higher I trip across the growl of a pile-up, each operator frantically calling. Within a moment I realize that whomever they are calling is transmitting somewhere else, as the stations calling seem to be spread out over the better part of ten kilohertz. If the DX station being called was working stations on frequency, everyone would be calling virtually dead zero with each other. Hah! There. Some one scored. "Roger, Rashid. Thanks for the contact. You're five by eight, five by eight, here in Alabama, Alabama is the QTH. The name here is Fred, Fox Romeo, Echo, Delta, Fred. Back to you. This is Whiskey Four November Kilo Italy, over."

Darn. The Four didn't give his contact's call. But the DX station's name is Rashid? Certainly likely to be a Middle East name. Needless to say, my interest is whetted. But where is the station? I consider for a moment. I was tuning higher up the band. And most DX stations working split have stations call them higher in frequency. Which means that if that is true here I must have tuned across the station already without noticing him.

But I was tuning carefully, so I really shouldn't have

missed him. Of course there's no law that says he can't have callers transmitting below his frequency. Not common, but surely not impossible either. Hmm. Where am I. OK, fourteen one eighty. Say, maybe he's transmitting below one fifty. Yeah, that's a thought. I started tuning at one fifty, so he could be lower than that and working split. Yeah, that's the most likely possibility. I'll dip down and have a look...

I punch the synch button on the VFO to lock on W4NKI's frequency, then switch to the other VFO and push it down to 14,150. I decide to tune down from there, rather than going lower and then tuning back higher, as the DX station is more likely perched close to the edge of the Stateside phone allocation than to be way down in the band. I flip momentarily to the other VFO. I hear a lot of stations calling, so presumably the DX station is not transmitting yet. I tune very slowly, every few seconds flipping to the pileup area. Ah, there, the pile-up is dying down rapidly. I flip back to the other VFO and start tuning more quickly now, knowing that the DX station is very probably transmitting, working some one.

Wait, what's that? ... "Name here is Rashid, Rashid. And now the microphone to you, my friend. Kilowatt Four Canada Echo Fox from Yankee Kilo One Alpha Alpha, over." Hah! It's the legendary YK1AA, on 14,145. I should have known. Only one problem, though. His beautiful and distinctive QSL already hangs within a few feet of my disappointed eye.

I switch VFO's and start tuning higher up past the pile-up on the YK. Well, at least I know the longpath into the Mid-East is certainly open. Rashid's signal proves it. A few kilohertz higher I trip over an LZ, weak, milking the long path as well. Just above him I hear a strong American voice, "located here in Dhahran. The handle here is Ken, Kilowatt England Norway, Ken..." OK, that's got to be HZ1AB. Bob, 9NZM, did say he was coming through in good order. I keep turning the knob.

A ZC4 entertains a small but avid following. I call that one in on the two meter DX net, then continue tuning. Several kilohertz above him I hear a CQ, but as the operator signs his call QRM from stations calling the ZC4 wash over the frequency. By the time the QRM finishes, the frequency is quiet again. Could it have been the 4W or the YI? Wait, there he goes again. Uh, QRM again. Wait, now it's clearing. Oh, OK, right. Darn. It's an A9 in Bahrain. Again, a nice one, but another I don't need. I call that one in on the DX repeater as well.

I glance at the frequency readout; OK, I'm at 14,223. I'll tune up to 280, then go back down. Above 14,230 good DX stations tend to be much fewer, but on the other hand when you do find something good the pile-ups are less severe. Also, because of this, less experienced DX station operators willing to work "W" stations tend to hide higher up in the band to avoid the massive pile-ups that can form in the lower part of the DX phone band. The same is true in the CW bands as well, with rare DX sometimes being unearthed by the patient operator even in the novice bands.

I keep moving slowly higher, watching every signal I hear. One thing nice about tuning for DX on SSB; the accents of the DX station can be a guide to their origin. Of course, one can get fooled that way; there are plenty of American and British operators operating from other countries, including some rare ones, and to pass over a signal because the accent is familiar can be a costly mistake.

As I muse, I keep tuning. And then, at 14,243, I come across him... "CQ CQ, Hallo CQ North America, Yemen calling, 4WØ Kilo Portugal, Four Washington Zero Kilo Papa, calling CQ..." My heart skips a beat. Good, clean S-6 signal, a bit of an accent. Hope no one hears him. Quickly I scan the gear. Yes, the antenna is dead on the long path into the Middle East. I am set up for transceive operation on the main VFO. Mic gain where it belongs. There, he signs. I pause a moment, then hear another station begin calling, apparently having paused as well. The other station sounds fairly strong, but with a lot of back scatter. But now I have the advantage. I know another station is calling, so if I make my call a few seconds longer longer than usual, I have a better chance of getting one clear call without QRM... "Washington Zero Kilo Papa, here is W9KNI, Whiskey Nine Kilo November India, that's W9KNI, Whiskey Niner Kilowatt November India, W9KNI, Over."

I listen. Hah. Some advantage... "Your five by eight, five by..." Whoever did get him, it sure wasn't me. The 4W is already handing out a report. That's the disadvantage of the delayed call. If another station starts calling before you do, there is a very good chance that the DX operator will "lock" his ears on the first station he hears, ignoring other callers that start their call moments later. And that is equally true on both

CW and SSB... "Name here is Ahmed, Ahmed. Please QSL via my manager, Alpha Papa Two Japan Kilo, that's AP2JK. Over to you, my friend, W8JBI from 4WØKP, Over."

W8JBI. Huh. Well, at least I got beaten out by one of the old pro's of DX'ing. He comes back, and I listen. I try to figure if I should have done anything different. Other than not pausing, which was a gamble anyhow, I can't think what it would be. Except maybe keep my call shorter; obviously W8JBI made his call short. All I can do right now is hope no one else finds the 4W before I get through. I listen.

The 4W is soon signing clear. The instant he stops, I call, this time signing only my call. No need to use his call at all now that he knows he's being heard, "Whiskey Nine Kilo Norway Italy, W9KNI, over." Darn. I hear at least three stations signing their calls. But the 4W is in there. Maybe I got him?

"..report is 5 and 7, my name is Ahmed, Ahmed, back to you, Washington Three Hotel Hotel Germany, this is 4WØKP, Over." Huh. Not good. Being beaten out by a three when the path should be favoring me. But maybe nine land will get a peak a bit later. I glance at my watch. Oops. I've got ten minutes to go before I have to leave. I notice a sinking feeling in my stomach.

W3HHG finishes quickly, and I call again, again being brief. I listen, only to hear more stations calling. There, the 4W is coming back to someone, but the call is buried in the QRM... There, he's turning it back. OK, it's Bob, K9RHY. One of the serious DX'ers a suburb away. So the nines are starting to make it through. Great. And me with seven or eight minutes left, and a pileup growing by the minute.

K9RHY makes his QSO short, and again I call. I listen. The frequency has turned into a disaster area. I think I can hear the 4W, so I do not make another call. He almost finishes his transmission before the pileup of stations still calling him begins to thin out, but one last gasp call from some idiot covers the call of the station he turns it over to. The frequency is silent for a moment. Could it have been me? No. There, he's calling a 4. I listen. Yes, it's N4RR. Wait. He's in western Illinois; yes, I met him at the W9DXCC Banquet. So maybe we're getting our peak now.

"...Name is Ahmed, Ahmed. Please QSL to Alpha Papa Two Japan Kilo. I shall QRT after this QSO; the QRM is becoming very bad. Over to you, my friend. N4RR, this is

4WØKP, over." The QSO is soon over.

I wait another minute; stations are calling madly, but apparently the 4W meant what he said. He is not heard again. I put my tie on as I listen to the frantic calls, then turn the rig off and head for the garage. As I drive off, I realize that I am secretly relieved. By the time the 4W went QRT, it was plainly obvious to me that my chances of getting through in the pileup were slim at best. There were too many stations on the frequency, and if the 4W continued operating transceive the situation would have become hopelessly out of control. I leave the house in peace, knowing I'm not missing a chance to put him in the log. It wouldn't be much of a meeting with me sitting there thinking of all the people getting through to the 4W while I'm locked up in a conference room.

And I learned a lot about his operating. He dislikes pileups, apparently lacks the equipment to operate split frequency, is not above operating higher in the band, locks his attention on the first station he hears, and is a reasonably competent operator. I've noted these points in the margin of my logbook, not that I'll easily forget them. And I have a good idea of what his voice sounds like, in case I hear him again in mid-transmission.

What I have learned doesn't sound like much, but it will be of considerable value in hunting him again, and I'll be ready to move quickly when I do find him. Assuming he is for real, as everyone seems to think he is, I'll certainly be putting in the time hunting for him. After all, what choice do I have? I'm a DX'er.

THE COMPLETE DX'ER

13 Summary, Section One

At this point you have finished Section One of this book. If you started out as a novice DX'er, or a DX'er trying to improve your effectiveness, you will have learned a great deal, especially if you have been reading this book a chapter at a time. And, your country worked total should be considerably improved.

By now, you will have figured out several implicit lessons "written between the lines." One of course is that the more time you put in listening, the more countries you are going to work. But remember, you have the rest of your life to work DX. You can't wait that long to meet your current obligations to your family or job.

Another lesson is that a good DX'er is always learning; learning about propagation, about radio, about people in other parts of the world, pile-up techniques, and more. And that should be a major part of the joy of the hunt.

The purpose of the first part of the book was to teach the fundamental skills that the successful DX'er should possess. Certainly, you will meet DX'ers with high country totals who lack many of the basic skills; it can be done. But it is a lot easier for a skilled operator.

Section II of this book is based on the belief that the reader is pretty well in command of the skills of DX'ing, and has worked something over two hundred countries. Finding the new ones gets harder and harder, with the time between DX'peditions to otherwise inactive countries still needed longer and longer. Now, we start the hunt for the elusive ones that are on the air, but rarely if ever found out in the open dealing with the pile-ups.

We will discuss other aspects of DX'ing besides operating, such as QSL'ing, towers and tower ordinances, the equipment philosophy of the DX'er, and TVI. We will discuss DX bulletins, buddy systems, HF DX nets and two meter DX nets. And a very important but brief chapter will cover DX'ing morality.

And, there will be a very short Section III, advice for the DX'er who has 300 countries, has worked everything that's on the air, and is waiting for the DX'peditions he needs to make it to the Honor Roll - the goal of the DX'er.

Good DX'ing!

14
QSL'ing

The old and oft abused saying has it that the job is never done until the paperwork is finished. And, for the DX'er, a country can't be safely scratched from the need list 'till the QSL is on the wall; and indeed has been accepted by the ARRL DXCC desk. But getting cards from all the stations you work is a difficult task; one never fully successful. And different techniques must be used for different countries.

Of course, the cheapest way to QSL is to utilize the outgoing bureau for transmittal, should your national society have one, and the corresponding incoming bureau for returns. The bureau system is rather reliable, thanks to the efforts of the national societies and dedicated volunteer work by our brother amateurs, and even some non-amateurs.

However, the bureau can be excruciatingly slow for the DX'er intent on running up a score. And, the considerable delays often imposed by the bureau system leave the DX'er questioning if the DX station he worked in some certain needed country is genuine, and if so, whether or not he QSL's at all. After two or three years without response, the DX'er does begin to wonder.

Often, the rarer DX station uses the services of a QSL manager, who assumes the responsibility of exchanging QSL's with all applicants in the name of the DX station after he has received the DX station's log books.

The third route generally used in QSL'ing is the direct QSL, sent via air or surface mail, and with or without provision for return postage in the form of currency, International Reply Coupons or a self addressed stamped envelope (commonly referred to as an SASE).

First, let's discuss your own QSL card. There are many QSL printers offering semi-custom QSL's for prices that range from exceptionally reasonable to outrageous. And, quality is not necessarily a function of price. The card should be attractive, easy to read, and should have the QSO data on the same side as your callsign. This may sound unnecessary, but there is little question that you will get more cards back with your call made out wrong if you have the data on the opposite side from your callsign. And receiving a card with your call filled out wrong from a country you need makes for a very unhappy day. Also, cards with all the data on one side are cheaper.

If you live in an area with any distinctive feature, it is nice to have a QSL reflecting this if the expense is not too high. Anything you can do on your QSL to make it more interesting and noticeable in a favorable way will help your QSL returns from the rare ones. But keep it in good taste!

When QSL'ing any station, exercise a little quality control on your end! Be sure the DX station's call is filled out properly, for starters. Be very careful that the log data entered on the QSL is correct, especially the date and time. Remember, in most locations at least part of the day has a different GMT date from your local date. Your chances of getting a QSL back decrease very rapidly if the time or date is wrong, especially if the QSO was made in a contest or with a DX'pedition. And if you really want to ruin your chances of getting a QSL, use local time on your QSL rather than GMT!

If you choose to send out your QSL's via the outgoing

bureau of your society, you will need to follow their instructions on transmittal of the cards.

As referenced above, many of the rarer DX stations utilize QSL managers to handle their QSLs. Often, the managers are American hams, making the obtaining of a QSL fast, easy and cheap for other Americans. (A high quality list of QSL managers for various DX stations is published monthly. Details on this list will be found at the end of this chapter.)

If you are sending your QSL to a manager for a new station, you should remember first of all that the manager himself is almost invariably a volunteer, and it is certainly in your interest and his to make his job as easy as possible. Always be sure to include a self addressed stamped envelope with your QSL, or, if the appropriate stamp is not available, International Reply Coupons (IRC's). On the back flap of your return envelope, you should write in complete QSO information - both your call sign and that of the DX station worked, date, time, mode, frequency and reports exchanged. This will help the manager sort the cards while he waits for the logs from the DX station.

Sometimes DX stations tell the whole world that some certain station is their manager and then neglect to send their logs to their manager. The great majority of managers are very responsible to the job, but if the DX stations fails to send the logs in, the manager can catch a lot of grief for a situation for which he has no control at all. Don't add to it! If, after no response for an extended period, you send a polite note with an SASE to the manager asking confirmation that he is holding your card, you will surely get a fast answer. But remember, amateur radio is the manager's hobby too. Don't ruin it for him.

Often, one reason managers handle QSL's is that they are stamp collectors. It never hurts to include a few unusual stamps, domestic or foreign, with your QSL to a manager. Another thought to remember is that very often the manager pays for the QSL printing for the DX station, not to mention other incidental expenses. If you are able to afford it, the occasional dollar slipped in the envelope to help defray printing expenses will surely not go unnoticed. But do include a note to tell the manager that is the reason why you made the inclusion; you do not wish to suggest impropriety.

As referenced above, many managers are American. Non-American stations chasing DX who utilize the services of American QSL managers on a regular basis can effect a substantial saving on return postage by getting an American amateur to purchase the appropriate U. S. stamps and mailing them to the non-American DX'er.

The aggressive DX'er generally, if he can afford it, sends his QSL's via air, with provision for return postage. Without careful attention to several details, this can be a frustrating and expensive exercise. The first problem, one unrecognized by many American DX'ers, is that often QSL's from DX stations are metric in size, and larger than the standard U. S. format. This means that the DX QSL may not fit the envelope that the American DX'er supplies.

The solution to this is to have a friendly overseas amateur send a packet of metric sized QSL envelopes. There are some rather nice blue air mail envelopes offered in most metric standard countries that will serve very nicely; and they are opaqued as well to help discourage theft. Alternately, such envelopes may be purchased from domestic suppliers. (See note at the end of this chapter.)

Which brings up the next major problem in QSL'ing; theft in the mails. Fortunately, the postal service in the majority of countries is reasonably honest, although there are certain countries where the only thing certain is that the mail will be rifled for any possible item of value. In any case, the wisest policy is to conceal return postage or any suggestion that there is return postage, money or IRC's from easy view.

This sounds easy. It is not. First, make up a dummy mailing to a DX station in your normal manner, using a blank QSL and an unsealed envelope. Now, take the mailing out into the noon sun, and hold it up. Remember that the sun in some parts of the world may be brighter. If you see any hint of an IRC, or even of a return envelope, you are already in trouble. Now, if you do pass that test, tap the side of the envelope sharply on a table, so that the internal contents of the envelope shift. Hold it up to the light again. Still OK?

There are several techniques that will help obscure the contents of the envelope. One is to be sure that all the contents of the envelope stay together. This can be done by taping things together, using small pieces of masking tape or removable plastic tape which can be removed without damaging the contents. Try pieces perhaps a quarter inch by a half inch (6X12mm). First, fold your return envelope, and lay it on the back of the QSL, being sure that it lies within the edges of the QSL card. If you are including IRC's, lay them over the folded envelope, blue side up, slightly staggered. Now, lay one of the pieces of masking tape across the corners of the IRC's, so that they are gently attached to the QSL, trapping the envelope beneath them.

The IRC's will help hide the return envelope, which otherwise might suggest to the potential thief the idea that there may be some form of return postage worth stealing. The tape pieces will keep things from drifting when tapped, making detection by the thief harder. The density of the whole packet inside the envelope will make it impossible to determine the contents.

There are variations that can be used to add security to your mail. One is to procure some brown or black plastic film and cut out pieces of it to opaque out the walls of the envelope. Another is to use carbon paper as a shield. It can be messy, but it is extremely black! But remember, getting too clever may in turn raise the interest of the thief.

There are other tricks. For example, it seems to be nearly impossible to mail QSL's with IRC's to Reunion Island from the States and have them get through. Apparently, while the mail is en route, at some point it is all opened. The trick there is to send the mail with a French address. In other words, use the complete Reunion address, especially including the postal code, and including the word "Reunion", but NOT the word "Island". Then, on the bottom line, write "France". The American post office then sends the mail to France, rather than directly to Reunion as it would otherwise. Reunion is a French department, and the mail is then forwarded to Reunion as domestic mail, which has a far better chance of making it through than mail directly to Reunion from the States.

The next question is how to make provision for the return postage so that the DX station can send you his coveted QSL, hopefully via air mail. From the States, two methods are usually used, IRC's and "green stamps", which are dollar bills. If IRC's are sent, it is necessary to know how many are required for an airmail return if desired. The easiest way to get this information is from the Second Op. It can also be obtained from the local post office, or the Callbook.

IRC cost the better part of a dollar each, and for an air mail return invariably at least two are required, often more. On the other hand, a dollar is almost always enough, converted to the local currency, to pay for return airmail. BUT, in some countries it is illegal to possess foreign currency, especially dollars. This is usually not because a country fears that its citizens are being subverted, but rather because many countries have a great need for hard currencies, and do not allow their citizens to privately hold foreign currencies.

In any case, and whatever the reason, in certain countries it is illegal, and often these are countries where mails are routinely opened and checked for contraband. If there is any question at all, don't send currency! You surely would not want to risk sending a ham to jail, or get him in trouble. At the least, you greatly endanger chances of getting a QSL you need. If there is any question at all about sending currency, ask the DX station. And if in doubt, DON'T DO IT.

A trick in use by a few DX'ers and that deserves more attention is to send the DX station a pre-addressed label rather than an envelope. Then, the DX station can return the QSL as an airmail post card at a lower postage rate, rather than the envelope rate.

One barrier to obtaining QSL's from any DX station can be determining the correct route for the card, even though you are willing to send your card via air mail with provision for return postage. Under normal circumstances, a little listening, or a direct question to the DX station will reveal the route the DX station prefers for a QSL, be it a QSL manager or his direct address, or an indication that the "CBA" (Callbook address) is correct.

However, sometimes the desired information just is not available from the DX station, for example in a contest situation, and other means of determining the route must be found. There are a number of possible routes. The Callbook is always an important source of information, for QSL addresses and many other tid-bits of information, and should be in every DXer's shack. The excellent DX columns in QST and CQ magazines are a possibility, and a check in the last year's issues often will be profitable. Various DX bulletins, which we will discuss later, are regularly a good source of QSL information. There is also a QSL manager's directory offered which can be

THE COMPLETE DX'ER   117

consulted.

However, you will discover that some QSL routes result in your QSL being returned undelivered. Other routes seem to be QSL gobblers, with never a return, and no hint of whether your QSL ever got to the DX station. And, some DX stations just don't QSL at all, while others don't QSL reliably.

It is important to keep adequate records in QSL'ing. If one route does not work after a reasonable period, something else must be tried. The easiest way to keep the needed records is to make notation in your log book of the date the QSL was sent and the route or address used. The date of the QSO should be noted in your need list, so that you can access the log data quickly.

If you want the QSL from a particular station badly enough, there is usually some way to pry one loose. First, of course, you should have some reasonable idea that the DX station has actually received your QSL and is not responding before you start trying other techniques. In many places, it is often possible to find some one willing to hand deliver a QSL; a friend of a friend, or a willing tourist, for example. In other places, this is impossible.

What should be avoided is making too strong an overture to the DX station. If you keep sending dollars to some non-cooperative DX station, for example, in fact you give him no reason to QSL, because once he does, he knows that you will stop sending him dollars.

Another possibility is that the DX station does not have any QSL's. In this case, you should make out a QSL on an index card or similar, filling it out completely, but leaving a place for the DX station to sign it. Such a card, when signed by the DX operator, will be accepted by the DXCC department if they believe the signature to be genuine.

Sending your return envelope with proper mint postage stamps of the DX station's country is a good idea when possible. Such stamps can often be purchased from a stamp dealer, or a DX'ers stamp service. Remember, however, that such moves are generally reserved for hard case situations, and often in such cases nothing will work. But if you can't find a more cooperative station in the same country to work, you have little choice but to pursue these avenues.

QSL'ing is as much a part of DX'ing as tuning the receiver is, if your goal is getting to the Honor Roll. Careful and prompt transmittal of the card after the QSO will help you towards that goal. And the thrill of getting a new card in the mail is almost as much fun as working the station. And though you are convinced otherwise, it is very unlikely that your mailman is stealing your QSL's for the stamps!

Note: The W6GO/K6HHD QSL Manager List is published monthly, and is an invaluable tool for the DX'er on the way up. A sample copy may be obtained by sending a business sized self addressed envelope with postage for 2 ounces, or 2 IRC's for international air mail return, to P. O. Box 700, Rio Linda, California 95673, USA.

Note: A DX QSL'ing supply service is offered by WV4V, DX QSL Associates, 434 Blair Road NW, Vienna, VA 22180. Send SASE for prices and details.

15
Night Call

The phone rang in the darkness. "Huh," I muttered, or something to that effect. Bleary eyed, I looked at the clock, just as the second ring started. It read back to me its dreadful message "1:35." In the morning. I sat up sharply. Phone calls at one thirty in the morning mean one of two things. A relative or close friend has died - or there is a new one on!

"Hello?"

"Bob! Get on quick. 3Y2LA is on from Bouvet. He's on fourteen one four five point six, listening two hundred and up. He's S-6 out here." It was my friend Jim in the Los Angeles area, which is what he meant about "out here."

"Right. But can he be for real? There were no reports or announcements..."

"Yeah. But he looks pretty good. Says the operation is only going to be on for twenty four to thirty hours. Says he is part of a group of scientists going from Norway to the Antarctic for some geological research. He said their ship is a Norwegian Navy ship, and that they had planned to go near Bouvet, and if the weather was good land a small party using the ship's helicopter. But they weren't sure, so they didn't make any announcements, and he just popped up. By the way, he's not a super operator."

"OK. I'm on my way. I'll call you a little later. Thanks."

"Roger."

Within moments, I am bundled in my bathrobe and slippers, and padding quickly down the stairs towards the shack. Bouvet! A frozen speck of land, a lonely mountain really, sticking out of the frozen far South Atlantic. Operations from Bouvet have been exceedingly rare over the years, due to its isolation and the difficulties of getting there. I have seen photos of the island; a starkly beautiful place that ran shivers down my spine. And one I need; a really tough one, one for which no realistic hope of a DX'pedition had surfaced in years.

Twenty four to thirty hours. On SSB. With an operator not exactly world- class. And me with only a moderately competitive signal in the SSB pileups. My stomach twinged. I turn the main power switch on, turn the linear switch on, then grab my Second Op and dial out the beam heading. Right, 135 degrees. That thing sure is handy in the heat of battle. I swing the antenna down, and as the rotor grinds its way south from the trans-polar path it had been left on the night before, I put the headphones and tune the transceiver to 14,145.6. Almost immediately I hear him; perhaps S-4, wait, there, he's lifting a bit more as the antenna turns. Yes, there, speaking English with a Scandinavian accent. "Roger, QSL my friend. Thanks and 73. QRZ please, listening fourteen two hundred and up, this is Three Yankee Two Lima Alpha now listening." Oh boy. If he is going to go through all that rigmarole, this could be a long night. My stomach rolls again as the shack phone rings.

It's Jim again. "Bob. I just got him. I was transmitting on two-thirty- six. He generally seems to be moving up the band about a kilohertz each contact, but sometimes he doesn't move, and sometimes he seems to just jerk his receiver dial and land anywhere."

"OK, Congratulations. Hope he's real. And I hope I get him."

THE COMPLETE DX'ER 121

"Yeah. Good luck. The pile-up is growing pretty fast."

"Thanks." I put the phone on the hook, and set the linear tuning from the numbers chart for 14,250. I turn the speech processor on, and move the microphone to the center of the table. I'm as ready as I'm going to be. I listen. Yes, he's on. "Kilowatt Five Foxtrot Alpha, OK, OM, I have you now, you are five and six, five and six. 73 now. QRZ, listening fourteen two hundred and up, this is Three Yankee Two Lima Alpha listening." I set my transmit frequency on 237 and call. Even with Jim's information, I'm probably not very close to the last station he worked, but you have to start somewhere. The needles on the linear and the output meter all flicker as desired, indicating that everything is working properly.

I end my call and listen. I hear nothing for three or four seconds, then he comes back. "The Kilowatt Four station, go again please." Mentally, I groan. I know from sad experience that a call that selective means that everyone in the pileup will continue calling, regardless of the prefix or number in his call. The 3Y would have been a lot better to just keep listening until he got a complete call, even if from some one else. There, he transmits. "Sorry, the K4, I can not copy you. QRZ? This is 3Y2LA, listening fourteen two hundred and up, over."

I call again, still really with no idea of what frequency to be on. I listen. There. He's coming back to some one. "W6 Canada Tokyo, here's 3Y2LA. You are five and five, five and five. Do you copy? Here is 3Y2LA, over."

I punch the transceiver switch and listen using the other VFO. Frantically, I tune, looking for the six. Yes, there, I hear a station speaking slowly... "You are also five by five, five by five, thanks for the new one, QSL? 3Y2LA from W6CT." I flip to the other VFO, and as the 3Y2 comes back to the six, I move my transmit frequency a kilohertz higher than the W6's frequency..

"Roger, thank you for the QSO. QSL via the Norwegian Radio Relay League. QRZ now, listening fourteen two hundred up..."

I call again, now with greater confidence that I am on a frequency where I have a good chance to be heard. I listen. I hear the last of a W6's call on the 3Y2's frequency, some poor fellow who got his switch position wrong. Immediately, several policemen respond with the usual "gentle" manners of the twenty meter DX fraternity. Which all adds up to obliterating the 3Y2's response. I hear him for a few moments, but copy is hopeless. But then the frequency finally clears, just in time for me to hear, "The Whiskey Nine Norway India station, go ahead, from 3Y2LA, over."

I grab the mike. "3Y2LA, here's Whiskey Nine Kilowatt Norway India, that's Kilowatt Norway India, W9KNI. Thank you old man. You are five by five, five by five, QSL? From Whiskey Nine Kilowatt Norway India, W9KNI, over." I listen, anxious that he got my call correctly. I groan with dismay to find his frequency covered by two policemen telling each other to remove various parts of their anatomy from the frequency. After a few seconds, they quiet down, just in time for me to hear the 3Y2 going QRZ.

I note the time down in the log, then enter the rest of the QSO, along with a big question mark that already is causing an ache in my stomach. The age-old question of the DX'er, indeed, the nightmare, runs through my mind. Am I in the log?

I try to think out the situation rationally. He DID have the second and third letters of my call correctly. But then, it was some seconds after I finished my transmission before I heard the QRZ; the timing was not as sharp as I wish it would have been. And he showed no patience in trying to dig out a complete call earlier when he dumped the Kilo Four that had called.

I think about calling Jim; it's already after midnight in California, but perhaps he's still up. I hope so... The phone rings once, and Jim is on the line. "Jim! Glad you are still up. Did you hear my QSO?"

"Only part. One of those clowns on the 3Y's frequency is a local, and he really covered the 3Y here."

"Tell me about it. But did you hear if he ever got my call correctly?"

"No, sure didn't. I knew that's what you were calling about. But I sure can't tell you. He might have, I just can't say."

"Ughhh. That was probably my easy shot."

"Yeah. Oh. Just before I called you, he gave some guy hell for trying to get a second QSO. Not that I blame the 3Y; the station he was chewing out had had a clean, clear QSO ten minutes before. Still, maybe he had a burst of local QRN or something and never heard the end of it. And it's not as if the 3Y2 is that super about picking out calls and making sure he's got it right."

"Yup. Do you think he's real?"

"It looks pretty good. One of the locals just said on two meters that his English DX Bulletin that came today - I mean yesterday now - anyhow it had a flash that a Bouvet operation was a real possibility, and that it was to be by a non-DX oriented amateur, and would be for a maximum of 36 hours, maybe less."

"So it looks to be real then."

"Yeah."

"How long do you think they have been on?"

"Not over an hour, probably less. It got called in on two meters just as I was going to bed."

"What do you think I ought to do?"

"Log the one you got, wait an hour, then try like crazy for an insurance QSO. Then write a letter with your QSL explaining why you might be in the log twice. 'Cuz I don't think you'll get a shot on another band. That English bulletin apparently said twenty meters only."

"Wonderful. But you're right. That's what I'm going to have to do."

"You got to work tomorrow? I mean today?"

"I did. But I'm going to take a vacation day unless I get lucky. I really need this one."

"Right. Good luck."

"OK, and congratulations on your QSO."

"Yeah. It was a real solid one. Good night."

I hang up with Jim, then get on the phone calling the local DX'ers who need Bouvet. The list is long. When I have finished that, I go get a cup of coffee, then return to listen before my self imposed quiet hour sentence is finished.

The situation has deteriorated. There are a number of stations around the 3Y's frequency, offering comments, telling each other to QRT, using various combinations of less polite words, etc. One station plaintively asks the world at large if someone can't take a list. That becomes the occasion of a number of DX'ers expressing their opinion of lists. The 3Y, apparently oblivious to all the traffic on his frequency, goes along placidly, leaving behind him what must undoubtedly be a long trail of broken QSO's.

I go for a second cup of coffee and return. I try to study the fellow's operating tactics, such as they are, in hopes of learning something that will give me an edge. But instead, my stomach sinks further as I listen, as the frequency and the operation get more and more out of control. And, as the hour gets later and I sit there sipping my coffee, I notice that the 3Y's signal is fading slowly.

I surely can't blame the 3Y operator for being a poor operator. The trip to Bouvet was not intended as a DX'pedition at all, and we DX'ers should consider ourselves fortunate that anyone at all got on. Besides, if all DX stations operated like OH2BH or DJ6SI or the like, every DX'er would be on the honor roll if he hung around long enough. But working this fellow for a solid QSO under the circumstances is going to be a nerve racking experience at best.

The situation on the 3Y's frequency finally begins to clear up. Even the jammer sending dits with the old bug gives up, and the 3Y becomes a lot more copiable. But now his signal is S-2 at best, with QSB taking some of his transmissions into the noise. Not only that, but skip seems to have gotten longer, and I can find very few stations calling him. Even so, Jim's comments about the 3Y's operating habits seem dead accurate.

The local DX'ers in the mean time have been busy calling the 3Y, with mixed results. Several of the big guns get through, most don't. Two meters is quiet except for the occasional call of the transmit frequency of some station in QSO with the 3Y; such spotting is the usual duty for those already in the log. But the calls become fewer and fewer as the skip goes out.

Then, the 3Y's signal improves slightly for a few minutes, just as he works a European, an LA station. He makes an announcement... "Gentlemen, the band now seems open to Europe, so I shall now work only European stations for the next few hours. Thank you to all the stations in North America. QRZ Europe, listening 14,200 and up, from 3Y2LA."

At least the fellow is a gentleman, and saves me needless grinding on the electric power meter. I empty my coffee cup, ask on two meters for a phone call from anyone if he becomes available again, and head for bed, first setting the alarm clock for 6 AM...

I lay in the bed, the excitement of the situation and the effect of the coffee denying me my sleep. I review the recent events in my mind. Clearly, I should have asked the 3Y2 to QSL my call during my transmission. But then he was giving both calls on his final transmission to me, or I must presume that he did; he did with every other station that he worked. Again, I THINK I worked him, but I can not be positive in my own mind. Was there anything else I could have done under the circumstances? Other than asking for a confirmation of my call, I can think of nothing.

I hate the thought of having to go for an insurance QSO on the same band; it is considered greedy by many, and some DX'peditions refuse to QSL stations that make a habit of that. But I really don't see that I have any choice. If I ever get another shot at him, I doubt there's any way he would remember my call anyway, unless he is using a computer AND I'm in the log correctly, in which case I can write a letter of fervent apology, and explain the circumstances. But the thought of trying to snag him out of what will surely be a monstrous pile-up churns my stomach. I can only pray that I'm already in the log.

It seems only a minute later that the alarm goes off. I must have managed to fall asleep after all. I rise, not feeling terribly terrific, and pad back down to the shack.

"Any sign of the 3Y? From W9KNI."

"Good morning Bob. Here's K9RHY. Yeah, he's still on, same frequency. But he's been doing nothing but work Europeans. He darn near faded out for a while, but now he's S-5 again. But thanks for that call last night. That surely was a goody for me."

"Glad you got him, Bob. But what are you doing up so early? You had a good solid QSO."

"I was too excited to sleep, so I've been doing homework in front of the rig. I'm going to go shower in a few minutes and get going."

"OK. How's the fellow doing?"

"Maybe a bit better. I think he's learning a bit. The QSO's are shorter now, and he's trying to make a better effort to see that he has a call sign correctly."

"OK, good. Well, I think I'll go get a cup of coffee, then come start my vigil again."

"Roger. I'll wait till you're back and watch him, then I'm out of here."

"Thanks. K9RHY from W9KNI."

A few minutes later I begin my vigil. The 3Y2 is good copy in spite of occasional catcalls and unkind comments on his frequency. But, as before, he ignores them, and steadily plows on. I do notice that at times the European stations seem to be having trouble copying him, but he works patiently and seems to be making a determined effort to get solid QSOs. My respect for the operator grows. He may be new to the game, but he seems to be learning, and is clearly trying to do the very best job he can. I keep listening.

After some number of QSO's, I hear him suddenly start working a station in another language; although I do not understand it, I recognize it as a Scandinavian tongue, doubtless either Norwegian or Swedish. Which of course makes sense. After several exchanges, he switches back to English. "This is 3Y2LA, now going QRT for a commercial schedule. I shall be back on this frequency in perhaps 20 minutes. Thank you. 3Y2LA QRT."

I glance at my watch; it's 7:13. Hmm. Strange time for a commercial schedule? One would expect such a schedule to come on the hour, or at least on the half hour. And he promises to be back in about 20 minutes. Interesting. And he just worked a station in Norwegian before he QRT'd. Hmmm. Betcha what he really did is sneak off somewhere and is working a bunch of Norwegians on some secret frequency. That does make sense. What to do? If I could find the frequency, maybe I could get in on the easy shot. Yeah. I leave one VFO on the 3Y's frequency, and quickly start tuning the band with the other. I decide to forget CW, and start tuning from 14,100 up...

Although I tune very carefully, both with the antenna on Bouvet, and then on Europe, I come up with nothing. The tuning turned into a tough assignment; if my thesis is correct, what I am looking for are stations making what are presumably brief transmissions in Norwegian, a language I am unfamiliar with. To do so means that one has to tune especially slowly. I did pop the bandwidth to the eight kilohertz position to give me more of an idea of what was on, but to no avail.

At 7:35 I return to the 3Y's normal frequency, get the antenna back on Bouvet, and sit listening, albeit impatiently. About five minutes later, I hear the 3Y return to the frequency, but, interestingly, about half a kilohertz higher in frequency. "I have returned. QRZ now, listening fourteen two hundred up, from 3Y2LA, Over."

Hah. No injunction against North American stations calling. I had set my transmit VFO right on 14,200 in just that hope. I call, then listen.

"K1MM, here's 3Y2LA. You are five by eight, do you copy? K1MM from 3Y2LA, over." I listen. Darn. K1MM was dead on the same frequency I was. But probably a better signal. And he got the QSO. Should I move up a kilohertz or stay on MM's frequency? I decide to slide up a kilohertz, and make ready to call.

"OK. Now, I listen for European stations for several more hours. QRZ Europe, from 3Y2LA." Darn! Well. I did guess right. And I was ready for the tiny window that did let one North American station through. But it wasn't me. I couldn't pull the trigger. I sigh, then go to call into the office to tell the boss that I am taking a vacation day.

I listen through the morning as the 3Y continues working Europeans. Then, later in the morning, he starts working Asian and African stations. As he goes, his skills at coping with the pile-ups seem to continue to improve slowly. I keep trying to track his receiving pattern, trying to develop an edge for when he starts taking North American stations again. I hope. But all I can learn is that Jim's description of the 3Y's tuning pattern seems to be completely accurate.

In mid-morning, he stops, and explains for several minutes what the trip is about, and states that he will continue to operate either about two or thirteen more hours; the length of time depending on whether the expedition decides to leave the island in late afternoon or early the next morning. He repeats the QSL information, states that he will work perhaps another hour of Asians and Africans, then take Central and South Americans plus the Pacific, then if there is time some more North Americans. My stomach grinds some more, and I wonder if I would damage any QSLs in soaking the stamps off in pursuit of what might be my next hobby.

None the less, I keep listening. My wife comes down to see how I'm doing, and I'm embarrassed to say I am rather peevish with her. "Look," she tells me, "I know you are all nervous. And you didn't get much sleep. Why don't you go take a shower and get dressed, then listen a few more minutes and if nothing is changed get out for a few minutes and take a brisk walk. It would do you a world of good." I recognize the value of her advice, and follow it.

When I return from my shower, the situation seems to be the same as before, so I go for a bit of a walk. It takes several minutes for the sharp late October breeze to blow away the fog I start in, but soon I am able to put the 3Y out of mind for a few minutes. The colors of a Midwest fall mixed with the invigorating breeze do wonders for my spirits. My wife is a smart lady.

Forty minutes later, I am back in front of the rig, refreshed, ready again to enter the fray. Nothing seems to have changed. But at 1820 Zulu, he starts a QSO in Norwegian again, one that lasts several minutes. Although he slipped into Norwegian on a number of brief QSOs, this time the conversation is longer. Then, he returns to English. "OK, my friends, please QRX for five minutes. I will return then." I wonder what's up, but I have an ominous feeling; he is nearly through the two hours he spoke of earlier. I sit and wait, but hardly patiently.

About three minutes later, he returns, but calling a Norwegian station in Norwegian. The station apparently comes right back. At this point of the day we seem to have pretty good propagation to Europe, and I can hear the LA station even off the side of my antenna. They QSO for several minutes, then the 3Y returns to English. "OK, here's 3Y2LA. We must QRT in ten minutes to prepare for our departure from Bouvet island. I will work all stations now, listening 14,200 and up. QRZ?"

My fears are confirmed. Desperately, I call, as I know literally thousands of others are doing, all despairing. Including me. All too soon, yet seemingly taking forever, the 3Y2, after four more QSO's, all with W4 stations, again addresses the hosts of DX'ers. "Gentlemen, and Ladies as well, I must QRT now. Thank you for the contacts. This is 3Y2LA on Bouvet, now going QRT." Stunned, I turn the rig off, put the headphones down, sit silently for a minute, then head for the stairs. All I can think of is my fervent hope that my original QSO would hold up and be in the log.

I'm sitting in my arm chair in the den, reading the morning newspaper and trying to evade the gloom that seems settled over me, when suddenly I sit straight up. Of course. He still might be on! He is likely working a few last Norwegian stations before they really do pull the plug. That's what that last Norwegian QSO was probably about, and why he left the frequency for a couple minutes - to see how much time he had left. How long ago did I turn the rig off? It couldn't have been more than five minutes. I jump out of the chair and race back to the shack. There's just a chance.

As I negotiate the stairs another thought hits me. I couldn't find their private frequency before - yet when he had left the frequency, - allegedly for a commercial schedule - and later returned, his transmit frequency had moved a good part of a kilohertz. So he had moved the VFO while away from the frequency. But, assuming he was working his Norwegian buddies, how come I could never hear him? Another flash hits me - he wasn't on twenty meters! He must have been on either forty or fifteen meters. If so, which band? I decide I don't have a choice. If it was forty meters, I'm dead regardless. No way can I do any good if they are running his pals on 7060. But if they are on fifteen meters, that might be another story. Of course, this time they could be on twenty meters on some other frequency.

I decide to gamble, and turn the rig to fifteen meters, and start tuning. I quickly establish that the band is in pretty good shape, and move the linear to the band as well to be ready, just in case. Might as well be hung for a sheep as a goat. I hear no Europeans, but I hear South American stations working Europeans, and giving them good reports, strongly suggesting a path between Bouvet and Norway. I set the selectivity to eight kilohertz and start tuning at 21,200, moving up.

Just as I tune past 21,240 I hit pay dirt. I hear a voice speaking in Norwegian, which I am now beginning to recognize, and quickly have him tuned in, and switch the selectivity to the usual narrow SSB position. Yes, it's him! The 3Y! I recognize the voice. He seems to be chatting for a moment, rather than trying to run stations through. Then I hear him apparently signing calls, and saying two words in English, "Seventy Three."

I waste not a moment. "3Y2LA, 3Y2LA, here's W9KNI, Whiskey Nine Kilowatt Norway India, you are five by seven, five by seven, over."

"Stand by, the W9." My heart pounds. I wait. Perhaps ten

seconds later, he comes back on, says a few words in Norwegian, then calls, "W9KNI, Whiskey Nine Kilowatt Norway Italy, here's 3Y2LA. OK, OM, you are positively my last QSO. You are five by seven also. QSL?"

"3Y2LA from W9KNI, I QSL, and thank you so very much. My name is Bob. 73 from Chicago. 3Y2LA, from Whiskey Nine Kilowatt Norway Italy."

"Roger, Bob. You are in the log. I am Ragnar. The helicopter is waiting. We are leaving the island in ten minutes. 73. 3Y2LA is now QRT." And he was.

16 Graduate Hunting

I was a bit late getting home. Several problems at the office had required an after hours meeting to clear up, but a phone call to the coast had finally helped settle things rather neatly, so I went home with a clean slate and a clear conscience. Dinner was waiting for me when I got home, along with the latest copy of "DX'ers Tout Sheet."

The rest of the family had finished eating, so I was able to read the bulletin quietly. The front page news was rather unexciting to me; there were several reports of DX'peditions to places I don't need, and none of DX'peditions to places I do need. 'Twas ever thus; the places I need at this point seem rather short of Hilton Hotels and 747's. Like the Laccadives or

Albania.

Inside, I begin to do a little better. Although there were no news stories of interest to me, there were reported sightings of several stations I do need. 5U7AR was reported by two stations on fifteen CW; apparently back from leave in France. That's good news; there had been reports that he was not going to return to Niger, and they seem to have quit issuing licenses there. He is apparently the only station operational, and definitely a rare one.

I check the times, dates and frequencies; yes, typical; around noon on week-days, local time. He's a tough bird to nail, which is of course part of the reason why he's rare. He seems to mostly prefer working the Europeans, and at that time of day propagation is iffy at best for us into the central African area on fifteen, and non-existent on twenty and ten at this time of year. After I finish chow, I'll go to the shack and make notes on my blackboard.

There are notes as well on ZK3KP, another of the more frustrating DX catches I need. Not that it's his fault; it seems there is no commercial power on that island in the Tokelaus, and he is using solar batteries to charge a motorcycle battery, with which he is powering a 5 watt QRP rig to a vertical. He shows up occasionally on twenty meter CW somewhere between 0800 and 1100 Zulu; real night watch stuff, and with a very inconsistent operating pattern. Most weeks he is on at least once, the essence of nightmares, of lost sleep, of frustration.

But then nobody promised us a rose garden. Anybody can call himself a DX'er. And a lot of fellows who confine their activities to working DL's on Sunday afternoon consider themselves DX'ers too. But forcing yourself out of a warm bed at 3 AM local time on a Tuesday to check a band that is probably dead for a station who is probably not there separates the DX'ers from the Weekend Warriors. And, from the looks of it, if I want to work the ZK3, I'm going to have to earn my stripes as a DX'er.

I finish dinner and head down to the shack. I glance over my little blackboard at the notes there gleaned from stations heard but not worked, from the "DX'ers Tout Sheet," from tips passed on the local DX two meter net, and from a couple of the on-the-air bulletins. I already know its contents by heart; nothing hot for this hour. I chalk in the additional notes on the

ZK3, and start a box for data on the 5U7. Maybe I'll get lucky enough to distinguish an operating pattern that will allow me take a vacation day to lay for him with a reasonable hope of success.

I sure would like to figure out something for that ZK3; regular watching of the band for that fellow is going to be really rough on me in terms of lost sleep. Maybe I could recruit a few of the locals to watch for him on a rotating basis. Darn it all, if he would come on just an hour or so earlier my West Coast buddy could check for him before he goes to bed and call me when he finds him; Jim would certainly do that for me. But his earliest reported operating time so far is 0800 Zulu, and that's midnight on the coast and Jim has to work too. Jim knows I need that fellow; certainly he will keep an ear open for the ZK3, but I'm going to have to take other measures if I'm to have a real hope of success.

But in the end I decide to hold off another week or two and see if there are any more reports of his activity before trying to organize a rotating watch system. Might as well optimize chances as much as possible before we start to lose too much sleep.

I settle down in front of the receiver for a little tuning, but it soon becomes apparent to me that some kind of disturbance is in progress. DX signals are few, and certainly not from transpolar paths, but rather South Americans and a couple southern African stations. I have a look for the Antarctic Islands, but there seems to be no significant activity.

I decide it's a good night for a little maintenance on the linear. I disconnect all cables, then oil the blower bearings, clean the insides of the RF compartment, and wash the tube chimneys so that the infra-red heat can escape. Then I reconnect everything and check it out.

I come across YV1NX calling CQ, and give him a quick call. He comes right back, and we have a nice QSO, exchanging a few DX tips. Fergus has worked the ZK3, and gives me the time and frequency, another item for the blackboard, more data for the file. He needs a 5U7 too, and has no information for me, but he is very pleased to hear that 5U7AR is QRV again.

We ragchew a while longer, then I decide to retire. Signals even from YV seem to be fading; a good night to catch up on my reading. I pull the switches, and head up the stairs.

17
Antenna and Tower Notes

There is an old philosophy about antennas that goes, "If it survived the winter, it wasn't big enough!" This approach seems to be the guiding design philosophy for many would-be DX'ers. It is also the wrong philosophy.

If we presume your primary operating interest is not contests per se but rather the pursuit of new countries, an accurate view point is that DX'ing is a year round contest. This means that you expect to be on the air at any time during the year, looking for the new ones, be it by tuning or by trying to crack a pile-up. But, if your antenna is down, you can't be effectively chasing that new country.

An effective DX'er must be able to deliver a signal, preferably with gain, on any amateur frequency where the possibility of finding a new country exists, and at any time. You may be able to rule out forty meters and eighty meters, but on twenty meters and above you must be operational and ready at all times. This implies a high level of reliability, both in your equipment and in your antenna system. And don't forget, that includes the tower, rotor and feed line as well.

Considered from this perspective, many antenna installations are distinctly too ambitious. Crank-up towers designed for small tribanders are too often pressed into service carrying huge stacked monoband arrays. Big yagis are rotated by TV rotors possessing no hope of survival in a winter storm. Two inch fence post pipes are used for masts to hold large arrays ten feet over the top of the tower. And owners of such installations spend two thirds of every winter QRT waiting for spring to arrive so that they may assess the damages and begin to rebuild.

Of course, the reasons for this situation generally either reflect economic reality or the real estate available for antennas, and certainly these are genuine considerations which cannot be ignored. However, where a compromise must be made, it is too often made in favor of a stronger signal rather than sound engineering practice.

The reliability of antennas themselves is an important consideration, especially considering the environment they are to be erected in. The author's present antenna system for twenty, fifteen and ten meters is a four element triband yagi on a sixteen foot boom, up sixty feet. Seven feet above that is a "shorty-forty" two element forty meter yagi, rotated 90 degrees to avoid inter-action with fifteen meters on the tribander. The mast to support both antennas is 2 inches (51mm) diameter 1/4 inch wall (6 mm) chrome-moly steel. The rotator sits below the top of the tower on a shelf, and a thrust bearing carries the weight of the antenna array. It is rotated by one of the larger rotors available to the amateur community. The tower is steel, with 18 inch (45cm) across the sides, and features both a house bracket and guy wires. The guys are attached to guy brackets with torque bars. Needless to say, the whole installation is considered conservative by any engineering standard.

The author kept a two element quad up for eight years, and attained the Honor Roll with it. The quad was originally chosen for several reasons, all of which were then valid. One certainly was economic. A full-sized two element quad was not only reasonably inexpensive in itself, but the small wind area and low weight meant that a premium rotator would not be required.

Another aspect of the quad was that as a tribander no compromise was made, unlike the typical trapped triband yagis. Being full sized on each band insured maximum bandwidth and gain across the full amateur bands the quad was designed for. And the antenna was a delight to use in situations where rain static wiped out reception on yagis.

But nothing is perfect. The mortal enemy of the quad is winter, and especially winter ice storms. Though the quad used was one of the better designs, and was improved as experience dictated, during several winters the author was QRT on the higher DX bands for some months, awaiting warmer weather to repair broken wires, and was back on the air only due to heroic efforts on the first day the temperature rose above freezing. Finally, it was retired in favor of the new yagi along with a new rotator.

The yagi has worked out quite well, being one of the new broad-banded tribander designs. It has come close, at least, to meeting the manufacturer's claims, and has withstood five winters without significant failure. Static noise in the rain is now a problem where it was not with the quad, and the yagi is probably little better in gain than the quad. But, in a land of snowy winters and icy springs, it has survived. And this is the kind of reliability the active DX'er needs.

For a good deal of information on a proper tower installation, Unarco-Rohn of Peoria, Illinois has an excellent book offered for two dollars. The information is for their own products, but is universal in application, and required reading for any one interested in a tower installation.

Regardless of what kind of tower is erected, the amateur owes it to himself and his neighbors to erect an installation matched to or preferably exceeding the surface area ratings of the antennas to be used with it, based upon wind conditions of your area. Guy wire should be purchased with attention to the load ratings the tower will be called upon to carry. Generally the offerings of the local Radio and TV shop will be found to be rather sadly lacking. Thimbles and cable clamps should be of premium quality. Absolutely insist that turn-buckle eyes are forged and pierced, rather than made from formed and bent wire. A good friend lost a beautiful tower and antenna system because a turnbuckle bought at the local mass-merchandise hardware department pulled open during a wind storm. If you have trouble finding the good stuff, check the local farm supply outlet, a contractor's supply, or a marine supply store.

Careful attention to lightning protection and grounding should be a part of the final design. And, consideration should be given as well to seeing that the tower is child-proof for climbing. Consideration should be paid to guy wires that come down to the ground to avoid injury to those taking an evening stroll, which could well include you or members of your family. Some installations use a heavy steel post set in concrete to avoid this problem, so the wires do not come down to the ground. Others take a piece of garden hose and slip it over the guy, so that accidental contact with it will not be as dangerous, and the wire more easily seen.

Many amateurs use climbing belts for tower work that are obtained, honorably or otherwise, from the local utility company. Often, these belts hang mouldering on a nail till they are called into occasional service, at which time they are used as a life support system. Needless to say, this does not exactly constitute good practice.

Belts should only be stored so that they hang straight by their tongues. At least once each winter, they should be cleaned and oiled with a preparation of neat's-foot oil, and given a good inspection and the careful attention your life deserves. Remember, DX'ing and towers are inter-twined. Do it right. It costs a little more initially, but pays big dividends in the long run.

Another place that the DX'er should not skimp or take the quick way out is in the purchase of coaxial cable. There are many grades of coax available, mostly ranging from poor to awful. Enormous amounts of junk cable were manufactured for CB'ers, and are still being offered. The serious DX'er has only two real choices for feed lines, particularly for twenty meters and up.

One choice is to go to hard line coax. If the antenna is more than a couple hundred feet from the shack hard line coax is almost a must. There is no sense in spending big money for a linear amplifier, for example, and then leaving hundreds of watts in the coax. Nor does it make sense to extend the boom length of the yagi for an extra dB, only to leave several dB in the coax.

If your antenna is not so distant, conventional coax is fine for your feedline. But Caveat Emptor! Do not buy any current RG8 or RG8/U coax. Coax of recent manufacture carrying that designation is junk designed to enhance the profits of CB dealers. The only coax the DX'er should buy for outside runs is RG213/U. Belden coax is especially recommended, although there may be other brands of good quality. Good coax may not be easily available at the corner electronics emporium, but most amateur radio dealers carry the good stuff. Stay away from the foam dielectric types for outside use; life isn't long enough for the grief it can cause.

An interesting new coax, style number 9913, has come onto the market from Belden, and should be given consideration in installations not using hard line. This coax features superb low loss characteristics, approaching those of hard-line at a significantly lower cost, accepts standard UHF connectors, and is entirely suitable for HF runs of hundreds of feet. Type 9913 cable does not have the full flexibility of normal RG-213, so cannot be used on crank up towers or for the feed to the antenna itself, but can be used to the top of guyed towers, where a short run of regular coax can go to the antenna.

And, while on the subject of feedlines, another false economy is the use of "bargain" coax connectors. These connectors will generally handle your transmitted RF just fine. They also will often degrade receiving performance, help cause TVI instead of helping prevent it, and can cause IMD product deterioration on SSB. But you can save as much as fifty cents each over a good Amphenol connector.

The above comments should by no means be considered to be complete and comprehensive. They are offered only to suggest that there is a good deal to the subject of towers, antennas and feed lines, and the rising DX'er intent on a good installation should spend time studying the subject and the literature from suppliers. And, as in so many other fields, you get what you pay for. Reliability, integrity and safety should be the design goals of a tower and antenna installation, rather than maximum height and longest boom without regard to long term survival.

18 Hunting Continued

After dinner, I settle into the easy chair with a cup of coffee, and a chance to read my new copy of "The DX'ers Tout Sheet." The front page offers hope of a DX'pedition to Mount Athos in six or seven weeks, and there is a rumor as well of a serious operation from S.M.O.M. in Rome, which would be a really popular item if it comes off, though I don't need it. And apparently one of the OH fellows operating portable C9M in Mozambique has secured proper documentation for his operation, as the ARRL is now accepting his cards for DXCC. That is great news for some of the gang who have been sitting on C9M cards.

Page two has more of the details of the aborted DX'pedition to the Kermadecs. They did the right thing in not pushing ahead; though really they never should have even attempted the trip during the typhoon season.

Things start to improve for me on page three, with two more reports on ZK3KP. I take the bulletin into the basement shack, and note the times and frequencies on my bulletin board; then get the calendar out and start trying to plot his activity, using the notes on the board.

Let's see here. OK, he was on the fifth, the eighth, the fourteenth and the twenty first. Darn. There is the hint of a suggestion of a pattern there; but it doesn't hold. He was on the fifth and the eighth, that's three days apart. Then he is reported on the fourteenth; that's six days from the eighth, or two times three. Maybe he was on the eleventh too, and no one reported it to the bulletin? That's entirely possible, of course, or perhaps the bulletin didn't report it, feeling that there were enough reports on the ZK3. But then, even if I'm correct so far, the next report is on the twenty-first, a week later, instead of six days.

If he has a regular weekly schedule, then he should have been reported on the seventh and/or the fourteenth, but he wasn't. I check out the times that he was reported. If he was trying to keep any kind of a regular schedule, the operating times would likely, though not necessarily, be very close, especially remembering that he's using a motorcycle battery as his power source, and can't operate much over an hour at a time without a recharge from his solar cell array.

And-a-recharge-takes-three-days. AhhhhhhhHA! Hey! Now we're getting somewhere. Let's suppose that he was on on the eleventh, and wasn't reported. We know he was on the fourteenth; that was reported in the bulletin. Perhaps he was on again on the seventeenth, and again not reported. That would explain that part. But why did he then show up on the twenty-first instead of the twentieth?

I can think of a couple reasons. One, of course, was that he simply had something else better to do. But he operates so regularly on that third day, if I'm right, it would seem that he puts a high priority on it. Heck, he's probably bored to tears out there, and getting on the air might be the highlight of his week. I wonder how propagation was. I decide that there may be a clue in that; and I know how to find out. I pick up the two meter microphone. "Hey, W9RY W9RY, here's W9KNI W9KNI; are you around, Hal?" W9RY keeps detailed notes on the propagation reports from WWV.

I get lucky. "W9KNI here's W9RY; go ahead, Bob."

"Good evening, Hal. Say, could you give me the propagation numbers for the eleventh, the seventeenth and the twentieth of this month? Over."

"Yeah, hang on a second, Bob. I'm getting my book out. Right, here we go. OK. The eleventh was flux 110, A-19, K-4. The seventeenth was flux 136, A-7, K-2. And the twentieth was flux 124, A-29, K-5. You got that OK?"

"Roger. Thanks, Hal. Say, you need that ZK3 too, don't you? I'm trying to figure out a pattern, and I wanted to check out a theory."

"Yes, I sure do, Bob. If you figure out anything, do let me know, OK?"

"I sure will, Hal. If I think I have it close, would you be willing to work a shift on a watch? It probably would be middle of the night stuff."

"Yeah, don't I know it. But yes, I need that fellow pretty bad, too, so if you think you're on to something, give me a ring. I'll carry my share of the load. OK?"

"For sure, Hal. Thanks for the info. I think maybe I am. I'll get back to you. Thanks again. W9RY, here's W9KNI."

"Roger. OK, Bob. Glad that I could help. W9KNI, here's W9RY clear. 73's, Bob."

"73 Hal."

Yes, that would explain things. The eleventh had poor conditions, so that the fellow may well have been on, but conditions were bad enough that the States probably didn't have propagation to him. And conditions on the twentieth were worse, really pretty bad. Probably after a bit of tuning around the fellow went QRT to save his batteries for a day with better propagation. It's all very speculative, of course, but it does make sense.

But what would be a good time to be watching for him? I check over the reported sightings for data. They seem to be all over a field between 0800 and 1100 Zulu. I try to discern a pattern from the times that he was reported on, testing to see if there is any correlation between date and time, but it appears hopeless. If there is a correlation, it's beyond any logic I can detect. Probably he doesn't even keep a wrist watch out there. I wouldn't either. But that doesn't help me work a new one and still get some semblance of a night's sleep.

And, of course, I am dealing with random and incomplete data. I have no guarantee at all that the reporters to the DX'ers Tout Sheet necessarily heard the ZK3 from the moment he started operating any given evening. They may have heard him five minutes before he closed down, or heard his opening CQ, or anywhere in between. But I can't complain. As it is, I'm delighted I have enough data to work out what appears to be a valid theory on his operating activity at all.

In any case, the information published earlier states he can't operate much over an hour before his batteries go out and he has to recharge, which in turn does suggest his operating times are rather random within that 0800Z to 1100Z slot. That also means that any given night we are to patrol for him, we are going to have to cover at least three hours total to be sure of hearing him, if propagation holds for us.

Hmmm. Let's see who would likely be willing to take a turn in a watch for him. I get the club need list out, and have a look. OK, I find MM, NUD, QVB, J, RY, and myself. There's a couple others too, but past experience suggests they aren't serious enough about DX'ing to lose sleep looking for a new one. OK. If each would watch for an hour in the night, we could cover two nights, and, if we fail the first time around, we could repeat the watch a second time. If propagation holds, we're almost sure to get him, according to theory. Yes, it's definitely worth a try...

19
The DX'er and His Environment

"The ideal QTH is an island in the middle of a salt marsh atop a high plateau." from an address by Lew McCoy, W1ICP; Dayton, 1979

In the States, at least, most people move from house to house every few years, and even though DX'ers are in no other way average people, in that respect they often are. And, certainly, the DX'er does not exist that, when he moves, does not try to pick a new address with a quaint name like "Hilltop Ridge Lane," "High Vista Drive," "Telegraph Hill Road," "Overlook Avenue" or whatever. And this is not facetious, either. Open the Callbook to any page, and look at the street names listed; especially for the holders of the higher classes of licence. You will see the above is true.

Generally speaking, (but not always!) picking the highest point in three counties is a fairly good start to a great DX location. However, it usually isn't available. And if it is, often it is because it lacks refinements. Like roads, electricity and telephone service. Of course, in such cases the clever DX'er can work out alternate approaches to such simple needs, but other members of the family have been known to take a differing view on such matters.

Site selection, then, generally becomes an exercise in optimizing the possible. Certainly, high ground is always a desirable aspect of any site being considered. And in places like San Francisco, it is a simple exercise to determine what high ground is. But in places like the Chicago area, determining what QTH will be favorable is not always so easy.

The answer to this often lies in survey maps. Survey maps show elevations of terrain at any given point, and typically with a resolution of five feet in height. Learning to read these maps is not difficult, especially when comparing a map to the characteristics of a known location. In the States, these are usually available from the State Geological Survey Commission for most or all areas in the state. And survey maps are available in most other countries as well. A call to the city or county engineer should be helpful in finding a source for such maps. But it would probably be wise to not tell him the true purpose of your interest, as will be discussed later.

What are the characteristics of the ideal QTH that the DX'er should seek on such a map? To consider this, first we must digress for a moment, and consider the effects of terrain on a signal. To achieve the best gain in a given direction, the antenna should be located on the side of a hill facing the direction for which the gain is desired, rather than the top of the hill as most amateurs would expect. By being on the side of the hill, gain is derived from a radiation component reflected by that portion of the hill that goes up and away behind the antenna. Were the antenna on the top of the hill, that reflection component would be lost to undesired paths.

In practice, this factor would be rather more significant in the case of a dipole, rather than an array having forward gain, but it is a real consideration. Indeed, if the station were located on the top of a hill, and there were none of the normal constraints that most amateurs suffer, the ideal antenna farm might well consist of arrays scattered around the sides of the hill, all well below the crest, with switchable feed lines to select the optimum antenna.

In the world that is real for most DX'ers, however, such a plan is not possible. More likely to be real is either the possibility of being on top of a hill, or to be sitting astride a ridge. The top of the hill, all else being equal, is the preferred location for the DX'er, as the ground slopes down and away in all directions, yielding a reflection enhancement that can be worth between 3 and 6 dB on the DX paths, not to mention the simple benefit of the enhanced height. However, such hills are often rare, especially in the urban environment, with ridges a far more likely possibility.

Being astride a ridge, while not as desirable as being on a hill top, can make for an excellent DX'ing location. Signals propagated in the direction of the ridge spine will not be attenuated, while signals beamed in the sloping directions will receive a reflection enhancement.

In every case, the desirable feature to be sought is to have the ground sloping down and away from the antenna site in the direction to be optimized. All else being equal, it is desirable as well to have the ground sloping away be good moist conductive earth, to help optimize the ground image.

If a location that meets these requirements cannot be obtained, the DX'er should concentrate on finding a site that has no negative propagation aspects. Aspects to be avoided include significantly higher ground in any direction, or ground that slopes sharply up away from the antenna site in any direction likely to be important. The trouble with DX'ing is that sooner or later any direction becomes important. But, slightly depressed ground need not necessarily be ruled out, as long as there are no major obstructions in any significant direction. Depressed ground often displays ground conductivity that will overcome slight disadvantages of elevation.

Care should be taken to avoid high tension lines too close to the site. Such power lines can absorb significant amounts of RF, especially if they are within a hundred yards or so of the site. And noise radiation from them will plague the DX'er at least on an intermittent basis the rest of his life!

The classic work in the amateur literature on site selection is G6XN's work, "HF Antennas For All locations," published by the Radio Society of Great Britain, and widely available. It certainly is recommended reading for all DX'ers, although some of Mr. Moxon's work may be considered controversial on questions of what antennas to select in a given situation. But that is part of what makes antennas an interesting sub-hobby of amateur radio.

Having selected a potential site that meets the requirements of the DX'er and those of his family as well is only the beginning of the work the DX'er must do before the antenna erection party, or for that matter even before the finalization of the offer to buy. Careful attention to the legal aspects of a potential tower installation is absolutely mandatory.

There are generally two important possible limitations to antenna erection. One is the zoning ordinance governing the area, the other is the possibility of covenants and deed restrictions on the title that prohibit towers and similar antenna installations. While a good deal of the work to check out such potential problems can be done by yourself, it should be directed by a skilled attorney who is knowledgeable of the local situation.

Before you even get to the point of seriously considering a particular property, it is wise to determine what the potential problems are. For one thing, a town with no amateur antennas to be seen anywhere may be showing you an early warning sign. Towns with buried utilities often have particularly rigorous ordinances against towers. If you are looking in a suburban area, it is certainly wise to determine which towns are more favorable.

There are several ways to determine the local ordinance situation in a given town. Finding a local amateur is a start, where possible. But caution must be used. If he tells you that towers are no problem, that is a good sign, but certainly should not be considered as definitive. If he tells you that towers are absolutely out, that is a pretty good sign that you may be in trouble. But again, hams have been known to not welcome the prospect of a serious DX'er moving in down the street.

A trip to city hall can get a lot of answers, but caution is definitely advisable. Asking a question that never came up before may result in fast city council passage of an ordinance triggered by an unwise question. It is smarter to first try to see a copy of the ordinance at the local library, if possible. If not, ask to see a complete copy of the ordinance set without letting the clerk know of your specific interest.

Of course, a trip to the local DX club meeting, should one exist, can unearth a good deal of the information needed in a few hours about what the antenna situation is in the area. Also, don't forget, if a possible site is in an unincorporated area, the county board still has zoning jurisdiction, with some counties more restrictive than cities inside them.

Recent court decisions against restrictive local ordinances (But NOT deed restrictions or CC & R's) have generally been favorable to DX'ers, greatly helped by the FCC's pre-emption declaration widely known as PRB-1. Which means that a fight against restrictive ordinances will probably be successful, and may even win the attorney fees you incur.

But a fight against a town is never fun. It tends to rile the neighbors, and sensitize them to possible TVI problems, even when you are off the air. All else being anywhere close to being equal, you are far better off trying to buy a place with no potential problems, rather than a place where they are guaranteed.

When actually looking at houses, be wary of houses involved by deed with home-owners associations. Many are harmless beer and barbecue societies, but others see themselves as keepers of the eternal flame and defenders of the covenants you will be required to sign before you gain title to the house. If a home-owners association exists, you must be especially wary of possible covenants contrary to your interests, and of architectural review boards that can force you take a tower down with no recourse. Obtaining legal review of the situation is mandatory prior to making a binding offer on a property.

It is absolutely essential that you have your attorney review the title to a house, prior to making a binding offer, to insure that there are no restrictive covenants attached that preclude outside antennas. Finding a recently built house without such covenants can be very difficult, especially in the American west. If those covenants exist, you can be forced to remove your antennas, and you have no recourse. You will likely even be forced to pay attorney's fees for the neighbor calling for enforcement of the covenant. Be absolutely certain no deed restrictions exist, or your dream QTH can turn into a nightmare. DXCC can be attained using a dipole in the attic. But it sure isn't fun.

Once you move into the new dream QTH, you should apply for a building permit for your tower(s) prior to erection. Having determined earlier that there is no impediment, gaining the permit should be no problem. If you are told that no permit is needed, ask for a signed and dated letter from the mayor or the city engineer to that effect. Also, note in your log book the date of completion of the installation. Dated receipts for the purchase of the cement for the base will further strengthen your position. These procedures will help you prove that your antenna installation was legal when erected, so that it will be given grandfather protection should a more restrictive ordinance be passed at a later date.

Having secured the high ground you always dreamed of, and having planned the tower installation that will further enhance your signal, it is time to give serious consideration to lightning protection. It is not the intention of this book to be definitive about lightning protection, but rather to point out that in a serious DX QTH, lightning can be a hazard to life and property, not to mention the rig! Study the literature, and be sure that proper precautions are taken as a part of the tower installation and erection. Safety first!

After you have moved into the new QTH and have the antenna up, a little discretion is called for to avoid long term problems with the neighbors. Like, don't immediately go all out at full power in the nearest contest. The neighbors will certainly have noticed your tower and antennas, and will be apprehensive at first that all their stereos and TV's are at risk. It is better to meet their fears head on, offering to cooperate with possible interference problems, while at the same time taking it extremely easy for a few weeks, operating barefoot, avoiding prime-time football games, etc., except of course where a rare new country is on the line. And even then, if you think that you can make the QSO by operating barefoot, by all means do. The longer you go without your first TVI complaint, the easier it will be to handle it.

A significant problem, especially in a new QTH, can be interference from the power lines, making weak signal copy difficult or impossible. The causes of line noise are varied, ranging from untrimmed trees and bushes rubbing against a power line to arcing insulators. Power pole transformers are often blamed, but rarely are the cause of a problem.

Getting rid of a problem can be difficult. Power companies range from extremely cooperative to downright obnoxious in their attitude towards power line noise. But in the States, at least, they have a clear obligation to cure an offending noise when it occurs in their power lines. American DX'ers be glad; our British cousins have no legal protection at all against line noise.

If you have a problem the first step is to try to localize the source of the interference. Unfortunately, this is not always as easy as it seems. But a good first step is to do a little checking with your antenna, if it is rotatable. Note both the direction the interference peaks from, and also the direction where the noise nulls out. Then, go outside and look at right angles to the bearing where the noise is nulled, as this is usually the sharpest. With a portable radio in hand, go in the direction indicated, and try to further localize the interference. Remember that noise is transmitted down power lines as though they were antennas, complete with SWR peaks at various points. Once you think that you have isolated the noise source, it is very wise to walk several blocks up and down, checking the noise at each pole, trying to find a stronger noise source.

When you have optimized location of the noise source as much as possible, it is time to seek possible allies in the form of the people living adjacent to the problem, especially if they are not served by cable TV. An inquiry as to how TV reception on the lower channels will often bring surprising replies, like, "It's terrible. It's all the fault of that new ham that moved down the block!" Once you have that problem sorted out, you have a useful ally for calling the attention to the power company to the problem. For some reason, power companies seem more willing to clean up their lines to resolve a TV reception problem than they are for a DX'er, and a call from an irate viewer can work wonders without requiring you lift a further finger.

Should it be necessary to deal directly with the utility yourself, there are several techniques that are useful. The first one is to always get the correct name of each person that you talk to, and write down the date and time that you talked to them. And be sure that you have the name spelled correctly; do not hesitate to ask the spelling. Refer to the person you are talking with by name several times, so they become aware that you know their name. These precautions may seem silly at first, but they will do wonders at getting the necessary attention.

If you are told the interference is your problem, politely but firmly ask to speak immediately to their superior. Eventually, you will get in touch with somebody who understands what you are talking about, and who understands the company's responsibility to you. When you find that person, be polite, and be patient. Sometimes it takes weeks to arrange to have a problem checked out and resolved. Curry the favor of that individual, and secure his extension number and engrave it in your log book; he can be very important to your country total!

20 Hunting - Again

"Hello, John? It's Bob... Yeah, fine... No, I haven't gotten my card yet, but I'm hoping. Good for you, though... Right, I did too... Hey, the reason I called; I think I've got a pattern figured out on the ZK3. The way I make it out, he normally is on every third day, because that's how long it takes him to get his battery charged from the solar cells. And if conditions are OK, I think we have a chance... Yeah, I know, but remember, that was at least four weeks ago, and we're a lot closer to summer conditions now."

"Anyhow, he's been reported on anywhere from 0800 Zulu to 1100 Zulu. Now, you need him, NUD needs him, QVB needs him, J needs him, RY needs him and I need him. That's six of us. So if we split it up and each watches an hour over two nights, and if we don't have any luck and do it again two nights later, we're almost sure to snag him. Are you willing to do it?... Yeah, I know. I agree, but he hasn't been reported outside that time frame at all, and if we're to get him, I think that's going to be the only way."

"See, the way I figure it, if he operates any earlier, he loses charging time, which means in the end he gets less time to operate. And he doesn't want that. So, if I'm right, that's going to be the only way that we're going to nail him. And the first fellow to stand the watch can call the second, and the second the third. That way, too, we know everyone held up his end. Or, if conditions are a wash-out, we can run it a night later, and give the first fellow a couple days off. I'll draw lots for the shifts if you're game. The way I have it figured from the "DX'ers Tout Sheet," we should start tomorrow night. I'll talk to the rest of the fellows, and draw the lots, and we'll finalize it on the net tonight. OK?... OK, good, see you later. 'Bye."

Ring... Ring... Ring... Ring...

Ugh. "Hello? OK, I'm up. Yah, I'll see you on two meters in five minutes. 'Bye." I groan.

My ever suffering wife remarks, a voice in the darkness, "Well, you asked for it."

"Unnhh. Yes. Don't remind me. But I'm sorry it woke you up too."

"Oh, that's OK. I always like to wake up for a few minutes at three AM, so I can go back to sleep and be thankful I'm not a DX'er."

I ease into my bathrobe and slippers, already laid out next to the bed in a futile hope I wouldn't wake my wife, and head down the stairs. I'd love a cup of coffee, but I'd probably lose any hope of sleep for the rest of the night if I did.

The basement shack is cold, and I'm glad my slippers have a lining to them. I switch on the main power switch, bringing the gear to life, and pick up the two meter microphone. "Hey, K9QVB, here's W9KNI. How's twenty?"

"Yeah, good morning Bob. W9KNI from K9QVB. Twenty's pretty good into the Pacific. The ZL's are banging in nice and loud, and so are the VK's, and there's an FK with a few people chasing him at fourteen oh thirty one. But no sign

THE COMPLETE DX'ER 155

of the ZK3, or anything else good."

"Right, OK, John. Thanks for the info and the call. I hope I'm waking you back up in fifteen minutes. Get off to bed now; I'll catch you here on two tomorrow evening. K9QVB, here's W9KNI."

"OK, Bob. Yes, that would be just fine, buddy. OK, I'm gone. Good night. W9KNI, here's K9QVB clear. Good hunting."

I look over the equipment, giving it all a quick check out. The antenna is on Africa, guess I'd better pull it around. My Second OP says 259 degrees from the mid-west for the Tokelaus; I start the rotator swinging the antenna around. I slip the headphones over my head, and start tuning. The ZK3 has always been reported between 030 and 045; I'll watch that area especially carefully, but I'll keep watch pretty well over the entire CW band. But first, let's have a look inside that zone.

I spin the dial to 030, and almost immediately come across a signal well over S-9, going along at a fine clip. Within moments, he turns it over to his contact, a VE7, OK, it's ZL3GQ. Hmm, the band can't be half bad for sure with signals like that. Which is good; while I have no direct reports on the ZK3's signal strength, it can't be anything too terrific if he's running ten watts of battery powered QRP to a simple antenna. But the band seems quiet tonight, and signals loud, so if he does come on, hearing him should be no problem.

I slowly move up the band, trying to get the feel of it. And it really is in lovely shape; not that it is open in many directions, but it does seem very nicely open to the west and the southwest. Signals stand out sharply against the quiet background hiss, free of QRN, like bright stars scintillate against a moonless sky.

There are weaker signals on the band as well; a P29 is just copiable, as well as a ZS4 trying to open a long path into Africa that isn't quite there yet. Or is he coming through short path, attenuated by the front to back ratio of my yagi? I could turn my antenna around to find out, but since I don't really care except for curiosity, I don't bother.

There are several VK's on the band, and a 3D2 calling CQ with no apparent takers. But I have other fish to fry; I tune on. It's sad that such fine conditions often occur when most DX'ers must be in bed. I hear the FK that John mentioned finishing a QSO, then, very weakly, several W stations calling him. He has a nice S-8 signal, while his callers are only just copiable. It sure is nice sometimes to have no short skip, though it can be very frustrating in pile-up conditions. But with conditions like they are tonight you know that virtually every station you hear is DX.

I glance at the local time clock; it's 3:20 AM. I still have forty minutes to go on my watch; then it's W9RY's turn. Funny how I'm so wide awake; I'll probably never get back to sleep. Oh well. If I feel that good, perhaps I'll keep Hal's watch; then we have an extra leg for another night. I must be crazy. But what do you expect? I'm a DX'er. And after all, you don't have to be crazy to be a DX'er, but it sure helps.

I slowly keep working my way up the band, till I'm at 080, when a teletype station reminds me I'm pretty high; I move the receiver back down to 030, and start climbing higher again, trying especially to dig out every weak signal, and see who it is. There's one, calling CQ. Hmmm. Oh, it's a K6. Let's see if anyone comes back. He's probably fishing for Europeans or somewhere where we don't have any propagation. There, he signs. Yes, there's a weak signal coming back...

Hey! It's him! It's the ZK3! I tear off one headphone, and jam the telephone hand set into one ear, listening to the frequency on the other. I'm punching K9QVB's number with one hand, trying to hold the phone with the other, while I wait for the K6 to have his QSO.

Hey! the K6 didn't come back to the ZK3! The frequency is quiet. I punch the synch button on the transceiver and grab for the paddle. I haven't got the linear running; the K6's failure to respond to the ZK3 caught me by surprise. QVB's phone is ringing, but I can't pay attention to the phone and send at the same time. I lay the phone handset on top of the keyer, turn the keyer monitor volume up, and start calling the ZK3. Maybe I can get him barefoot. As I call, I turn the gain up on the keyer, so that if when QVB answers he'll hear the CW, and know what's happening.

With bated breath, I listen. There, he's coming back. "QRZ QRZ W9? ? DE ZK3KP AGN K" As he starts the QRZ, I bang the linear power switch on, and watch the plate voltage needle come up. I quickly switch the RF watt meter into the high range, and put one hand on the drive control of the exciter. I start my call again with reduced drive, quickly

THE COMPLETE DX'ER

bringing it up till the linear is perking along with the correct amount of drive, "DE W9KNI W9KNI W9KNI KN".

Again, I listen, anxiously. Yes! There he goes, music to my ears, "W9KNI W9KNI DE ZK3KP R GE OC ES TNX..." I shove one earphone back, and pick up the phone again. But even as I do, the two meter speaker opens up, "OK, Bob, I'm here. What frequency? Here's K9QVB." I hang up the phone, then pick up the two meter mike, while I try to copy the ZK3.

"Fourteen-oh-three-seven. Call the rest of the fellows. I'm in QSO." Naturally, I'm terse, and probably not speaking too clearly, but no problem; John understands.

"Roger. Congratulations. I'm calling now."

As soon as I know QVB is OK, all my concentration is on the ZK3. And he IS pretty weak, only really just enough signal for adequate copy. But I get all the important things. He turns it back to me.

"R FB KEN ES TNX QSO ES NEW COUNTRY UR RST 549 549 HR NR CHICAGO NR CHICAGO ES NAME BOB BOB VY PLEASED TO QSO U AS 6 OF US BEEN HUNTING YOU IN SHIFTS HI HI GLAD CU WITH GUD CONDX HW CPY ZK3KP DE W9KNI KN". My longer than normal transmission helps give the rest of the lads time to assemble. I hear QVB talking to Hal on two meters, but copy of the ZK3 is not easy, and I turn the 2 meter rig down so that I have less QRM.

"R W9KNI DE ZK3KP R FB BOB ES TNX INFO YES CAN ONLY OP EVERY 3 DAYS WHEN BATTERIES CHARGED FM SOLAR PANEL SO GLD CONDX GUD ES THAT U FOUND ME BT PLS CALLS OF UR PALS QRV? RIG HR 10 WATTS TO GP HW CPY W9KNI DE ZK3KP KN".

"R ZK3KP DE W9KNI R FB KEN ES TNX DOPE ES PALS QRV ARE K9QVB K9QVB W9RY W9RY W9NUD W9NUD AI9J AI9J ES K9MM K9MM ES TNX KEN FOR QSO ES NEW ONE RIG HR KW ES 4 EL YAGI BT UR QRP FB HR OK KEN NOT KEEP U ES OTHERS QRX SO 73 ES TNX AGN ES GOOD DX ZK3KP DE W9KNI SK"

"R W9KNI DE ZK3KP FB BOB ES TNX VY MUCH FOR QSO GLAD TO QSO U ES UR PALS 73 QSL SURE SK NW K9QVB K9QVB DE ZK3KP KN"

Phew! I turn the two meter rig back up as I hear K9QVB calling the ZK3, and starting to give him a report. "Hey, Bob,

THE COMPLETE DX'ER

way to go. Nice piece of work there! Here's K9MM."

"Yeah, good work there, Bob. Looks like we're into a good one, from W9RY." And the rest of the fellows on frequency offer their congratulations as well, both for my getting through, and for the knowledge that they are next.

It's a good feeling; knowing that we've worked together to raise a rare new one. Winning as a team is fun in DX'ing too, especially when it's your pals, and all have worked and sacrificed to snare a real trophy.

And, working a station like the ZK3 is particularly satisfying in other ways as well. All the elements of an exciting game were there; early scouting reports, determining an operating pattern from incomplete data, developing a strategy, getting together a team, and winning the objective. Yes, logging the ZK3 is a particularly satisfying task.

I listen as the other fellows get their reports, and get into the ZK3's log. Under normal circumstances, a list type operation such as this would not be a very satisfying way to work a rare DX station, but in this case it is the culmination of a team effort, one that any of us might have initiated. And considering the hour and the long skip conditions, it is unlikely that there is much of a Stateside pile-up champing at the bit awaiting their crack at the ZK3.

The ZK3 finishes with the last of us, and after a few more minutes of mutual self congratulations, we break the impromptu net; after all, it is nearly four AM.

But I know I'm not yet ready to get back into bed; I'm too wide awake and still excited. I get a glass of milk, and tune for a few more minutes, while I try to compose myself.

The CW end of the band seems rather quiet, and I move the receiver up into the SSB reaches of the band. Signals seem to be getting weaker; often the case in the predawn hours before the band starts building towards the sunrise peak. But as I ease my way up a bit higher, I trip across a rather strong signal, one with a rather definite Australian accent. I pause a moment to listen, to see who he is; the other signals from VK land are now rather weaker.

"Righty-ho, then, mate. Now that we've established our base camp, we'll be gone off down the coast for about ten days, but I should be QRV again as soon as I get back. But I'm just not going to be able to be very active, we're going to be very busy while we're here, and three weeks really isn't enough time to do the job at all, but that's how it is. I'll be looking for you at the usual time and frequency, same as now. Dinner's called, I have to run right now. Give my love to Sylvia. VK6RS, here's VKØAH, Macquarie Island, off and clear and pulling the switch. Cheers, mate."

Hey! Macquarie - another one that I need! I do a panic QSY with the rig, and give him a quick, "VKØAH, here's W9KNI, W9KNI, you're five by eight, five by eight, go ahead," in hopes of snagging a quick QSO, but there is no response. He must have really meant it about going QRT, and it's obvious I'm not going to nail him tonight. And it's equally obvious that if I'm to have a chance to nail this one it's going to mean a lot of lost sleep and careful tuning. But a shot at Macquarie is a rare event, one I don't dare miss out on. It may be a long time before I get another shot. The season is is a bit unusual for an expedition to Macquarie, being nearly the dead of winter in that hemisphere, but governments everywhere have a logic of their own.

I make notes on the VKØ on my little blackboard, turn off the gear and head for the sack. It's obvious that I'd best start trying to get a bit ahead on my sleep; in ten days time I'm going to be needing all I can get. But what choice do I have? Any DX'er understands that.

21
Accessories for the DX'er

There are a number of accessories that will make the DX'ers life easier and more effective, and even more comfortable. Although some of the modern transceivers seem to feature every possible control and feature, including a good many that will never be used, there usually is still a need for a few items to make the station more effective.

The first actually is a technique, rather than an accessory. Headphones, which virtually ever DX'er use, are normally wired in phase, so that the apparent source of signals heard is outside of the head, and of course it is. But, reversing the phase of the headphones will make the apparent source of the signals seem to come from the middle of your head. This change, though subtle, is rather advantageous for copying CW in QRM and QRN, especially in weak signal conditions.

Reversing the phase of the headset is usually a rather simple procedure, requiring only the reversing of the leads going to one of the elements and leaving the other as is. In most modern headphones, a mini-speaker is used as the transducer. Usually, the side to reverse is the far side, the side away from the plug lead. Try it. It's well worth the effort.

An extremely valuable accessory that should be immediately at hand for every active DX'er is the Original N6RJ Second Op. This convenient reference offers quick answers to what are often urgent questions, such as beam headings, time differentials and ITU call block allocations. The Second Op has gone through numerous editions over the years to keep it updated, and remains a deservedly popular operating tool that should be in every shack.

A recently released Electronic Second Op computer program offers all features of the Original Second Op, while adding customized beam heading information for your exact location, daily sunrise and sunset times world wide, logging information, record keeping and an extensive list of other information. DX'ers using IBM PC's and compatibles will find this program invaluable. Another invaluable DX'ers program for the IBM PC world is "Terminator," a fascinating program detailing on screen a map of the world and showing the terminator line between light and darkness world wide, continuously updated for the date.

A similar aid for the astute DX'er lacking a computer is the DX Edge, which is a world wide sunlight/darkness calculator cleverly made from plastic. This interesting device utilizes a world map with different plastic overlays for each month, clearly showing at any given time which parts of the world are dark and which parts are light. A program duplicating the function is available for Commodore computer users.

The ability to change antennas quickly or to switch to a dummy load or a grounded position can be important to the DX'er for a variety of reasons, be it a panic band change in response to a call on two meters or a sudden lightning storm in the immediate vicinity of the shack requiring a fast disconnect. There are a number of switches available, both manually operated, which must be in the shack, and electrically operated switches which can mounted on the tower and operated remotely from the shack. Unfortunately, many of the switches offered to the amateur community are incapable of extended reliability, especially at full power.

The best electrically operated switches are the Transco style, but these are prohibitively priced unless they can be found in surplus or at a hamfest. However, if you are laying in a long run of copper hard line to your tower, and the choice is one remote switch or three feedlines, the price becomes more attractive. These types of switches utilize 28 volt control wiring, which makes for safer operation.

Of the mechanical switches as used in the shack, a series of switches offered by Alpha-Delta are an excellent choice, especially since these switches are well made and also offer partial protection against lightning.

A very useful addition for most receivers and transceivers is a good audio filter. A good filter will help eliminate white noise, both for CW and for SSB listening. It will help a great deal in digging weak signals out of the mud. And a good filter will reduce operator fatigue.

There are a number of filters on the market, some of which are superb, while others are not worth the money to buy. A new generation of switched capacitor audio filters is becoming available. A good example is the SuperSCAF filter from AFtronics. Every serious DX'er, whether CW or SSB oriented, should seriously consider the addition of such a filter. These filters typically utilize tunable cascaded low and high pass filters with extremely sharp response skirts. They will improve any transceiver's audio output, and can offer remarkable improvements in reception under difficult conditions in heavy QRM and QRN. What is especially notable about this new generation of filters are the adjustments allowed, the sharpness of the skirts, and the absolute lack of apparent ringing.

All audio filters offer variable selectivity, but some earlier models do so by continuously varying the "Q". Such filters should certainly be avoided, as this design is handicapped by severe ringing in the sharper selectivity positions, the very situations where the extra help is necessary.

At the time of this writing, other good filters are offered by Datong and Heath. Also, several of the MFJ filters are very usable, especially the ones offering switched bandwidths, rather than continuously variable ones.

An aspect too often ignored by the DX'er is the choice of microphone. Far too often, the selection criteria for a microphone is styling, cost, or "compatibility" with the transmitter - meaning that it comes from the same manufacturer. A moment's consideration should suggest that none of these criteria have any bearing on cracking pile-ups.

A DXer's microphone should NOT be chosen on the basis of fidelity or "natural" sound. The microphone should offer a cardiod pick-up pattern, and emphasize the high frequency components of the voice, rather than the low. The only real test of a microphone is its effectiveness in a pile-up. However, if possible, different microphones should be tried, perhaps in a contest situation, to develop a feel of which is best for the individual operator and his equipment. Certainly, this is hardly a definitive piece on the selection of a microphone, but if the DX'er catches the idea that some microphones are far better than others for DX work, he is on the right track.

A DX'er serious about the world he is working will be interested in a good atlas. Whenever the writer works an interesting DX station, he tries to find the exact location of the other station in an atlas. A paperback Penguin Atlas is kept on the shack table for quick reference, while two larger and much more comprehensive atlases are kept elsewhere in the house for further reference.

The problem with most of the more expensive atlases is that they always give enhanced coverage of the country of origin, and short shift to the rest of the world. It seems a bit inconsistent that the "All Universe" atlas has 65 double pages for the American States, and thirty-two single pages for the rest of the world, with an index to match. To test any atlas for quality and worth to the DX'er, try to find Pagalu Island (3CØ), also known as Annobon Island, Crozet Island (FT), and Bouvet Island, (3Y).

There are two atlases the author considers far and away superior to most of the offerings. The most comprehensive and complete atlas is the "Times Atlas of the World," Published by the Times of London. Though expensive, this is the most definitive work the author is aware of. Its biggest weakness is the lack of standard projection maps of the oceans, making location of some of the exotic island groups difficult to relate to other spots. In every other respect, it is a superb work, and the index is unsurpassed. It is available in better book stores.

The best value to be had for the money is the "National Geographic Atlas of the World." This atlas, while not quite as comprehensive as the Times of London Atlas, is not far below in quality in any way, and at rather less expense. And, it includes excellent maps of the oceans that the Times Atlas lacks. The National Geographic Atlas must be purchased through the National Geographic Society of Washington, DC. Either atlas is a superb investment.

For the CW man, the paddle and keyer are among the most important accessory items purchased or built. The paddle should be comfortable enough for the operator so that in the heat of battle over a new one he never has to give a thought about his paddle or his sending; it should be second nature. This of course requires practice, as well as a good paddle.

There are several important considerations to look for when choosing a paddle, be it iambic or single lever. For one, it should be heavy, so that it will stay put instead of having to be chased around the operating table. For another, there should be a definite tactile response to keying action. In other words, when you close one side of the paddle, there should be a click noise, and a feeling to the finger that the contact definitely closed. Some paddles have a rather mushy feeling when closed, leaving an uncertainty factor for the operator that slows his response and in turn his top speed.

Another important factor is a positive feedback response as you close a paddle contact. This is only available with paddles using spring return mechanisms. Some designs use magnetic attraction as the return force; however, the further the keying arm moves from the magnet as the contact is closed, the weaker the return force working on the arm, the force weakening on the basis of the inverse square law. Unfortunately, this weakening of the return force is exactly the opposite of the tactile response the operator needs from his paddle. A spring return, on the other hand, gives a stronger return force the further the contact arm is moved away from the rest point to the contact, and this is exactly the response the operator needs.

Gold plated silver contacts are very desirable on a paddle, as they will give the flawless closures required for the micro currents of modern keyers. Some of the alloyed metals, such as coin silver, will eventually corrode enough to cause problems of reliable closure for micro-circuitry.

The spring tension for the dot side and the dash side of a paddle should be separately adjustable, as most operators will have a differing preference for each side.

If a new operator is buying his first electronic keyer and paddle, he should definitely choose an iambic paddle. Iambic keying is a more efficient way to send a number of the letters of the alphabet, notably C, F, K, L, Q, R, and Y. People experienced in iambic sending are able to send at a faster and more comfortable rate, with less work, than operators on single lever paddles.

However, many amateurs have grown up using old bugs, and proceeded on to single lever paddles before the iambic technique was developed, and for these operators, the single lever paddle is still capable of excellent work. And any iambic keyer will work fine with any single lever paddle. But newcomers without the bug and single lever background should start right out with the iambic paddle.

Incidentally, the term "iambic" comes from the poetic usage of meter, or rhythm. A poetic metrical foot consists of an unaccented syllable followed by an accented syllable, and many types of poetry are made up of lines containing a number of metrical feet, such as iambic pentameter, which utilizes five metrical feet. This could be expressed as, "dit-dah-dit-dah-dit-dah-dit-dah-dit-dah" with unaccented syllables the dits and accented syllables the dahs. You will note that this is exactly the sound created by an iambic paddle and keyer when both sides are closed, the one alternating with the other, and hence the term "Iambic."

Fortunately for the novice today, virtually all keyers are fully iambic, whether the iambic feature is used or not. Only a few years ago, buying a keyer was an adventure, with so many different combinations of features and lack of features that selection was difficult. Today things are much simpler. Virtually every commercial keyer offered to the market is fully iambic.

Many DX'ers find keyers with programmable memory to be a very attractive feature, but their main utilization is in contest situations. They can be useful in pile-ups as well, especially for the operator who gets flustered easily. It can sometimes be easier to push a button correctly than a paddle! Other desirable features include variable weighting, dual polarity output, a built in monitor speaker and a minimum of front panel controls.

There are, of course, a great number of other accessories available for the DX'er, and many of them are virtual necessities in a well equipped shack. Examples include RF watt meters, speakers, 24 hour clocks kept on GMT time, a set of Callbooks, not to mention a good library of technical works, such as "The Radio Amateur's Handbook", Bill Orr's "Radio Handbook," and various antenna books by Bill Orr, "The ARRL Antenna Handbook" and RSGB's "HF Antennas in Their Environment" by Les Moxon, G6XN. Another book no DXer's shack should be without is Bill Nelson's "Interference Handbook."

Careful attention should be paid to keeping important spares available where it is economically feasible to do so. Murphy's Law guarantees that the surprise new country will become QRV for a short period while your rig is out of service. Spares include replacement tubes for your tube gear, possibly a new first RF transistor for solid state receivers, spare finals, and indeed, replacement equipment where possible. Many DX'ers, if they can afford to, do not sell off old equipment when buying a new rig, but rather keep the old gear as operating spares should the newer equipment fail.

Antennas, too, offer an opportunity for planning for disaster. For example, it is certainly a wise policy to have a trapped multi-band vertical stored away that can quickly be erected under harsh weather conditions the day after you've lost your HF antennas to an ice storm in mid-January, knowing that it will be March before the weather moderates enough to repair the damage. W6's, however, can invest that money elsewhere.

22
Special Language Techniques

"Hey, W9KNI, here's W9BW. Hey, Bob, was that you I heard parley-vousing with that French station? I didn't know you spoke French."

"W9BW, this is W9KNI. Yeah, good evening George. That was me. I'm pretty bad, though. I'm trying to learn at least a little ham French. Maybe enough to get lucky and nail one of the terrible 'T's. I have a suspicion at least a couple Africans I need are on, but only working in French to avoid big pile-ups, and I'm trying to learn enough to make the grade."

"Hey, good for you. Sounds like a good way to go after them. So how's the studying going?"

"Not too bad. I can muddle my way through a contact without chasing away the fellow at the other end. In fact, they almost seem eager to talk to me in French, even though I obviously know very little. At least no one has quit coming back to me. I started with a couple VE2's, and since then I've worked two or three F stations, and an FM. I'm trying to get more familiar with it before I go hunting bigger game. And I still don't have the alphabet down as well as I would like. It's a bit different in the pronunciation from English."

"Yeah. You wouldn't want to have TN8BK come back to you and you log him as TR8BK, haw haw."

"You got that right! And TN8BK is one of the stations I'm hoping to snag. I think he's the only fellow on from there, and I need that one bad."

"Yeah, I can understand that. I got an FQ8 station there right after they became independent. The fellow was running ten watts to a dipole. But it counted OK, and that's been a rare one ever since. Good luck with it anyhow. What are you using to study?"

"Well, I have a couple things. I've got my kid's freshman French textbook. And my buddy F2MA made a tape for me of the French pronunciation of the alphabet, and some simple QSO French as well, and I've been playing that over and over. So I think I'm nearly ready."

"Sounds good. But don't a lot of those fellows in Africa, especially the French speaking ones, hang out below our phone band?"

"Yeah, I think so. My plan is to listen down there and try to raise them on CW if that's where I find them. But maybe they work higher in the band too. And I'm going to try some CQ calls when I feel a bit more confident. I thought I'd try them at times we don't have propagation into Europe but we do into Africa. That may make it easier for me to cope."

"Yeah, right, you got to do something to keep those pile-ups down. Hey, it all sounds neat. But I have to run now. Good luck with it. W9KNI, here's W9BW."

"Roger. Fine George, I won't keep you. W9BW, here's W9KNI clear."

I put the two meter mike back down, and start tuning the bands again, thinking about that conversation, and my plan. I've got pretty well everything worked I've heard on the air, and the only chance I have of snagging a new one without waiting for a DX'pedition to hit is to try to smoke out a few of those Africans out. I need the TN8, a TJ1 and a 5V7. All are

French speaking countries, with mostly French speaking hams. While I don't KNOW there is activity from those countries except for the ever elusive TN8BK, it's my best shot for a new one right now. And, there may be another TN8 on besides Dr. Bernard, TN8BK. Maybe if I work hard at it I'll get lucky...

A cold, blustery January afternoon offers what may be ideal conditions for my quest. Twenty meters seems to close up tighter than a drum practically at the same time the sun goes down, even to the South Americans. By 7PM local the last few weeks, the band is a very large dummy load. That condition must be the same along and east of the Greenwich meridian, and the more so for the European stations further north, farther from the sun. That means the band on their north-south path is dead too around 7PM local time, probably even earlier.

But for stations nearer the sun bathed equator, twenty should still be open. But the path north to Europe is closed. So, the French speaking stations in central Africa wanting a QSO on twenty meters in the early evening hours local time are going to have to look for stations somewhere else to work. Like us. Because it's still early afternoon local time for us, and we still have good propagation, including a good path into central Africa...

"Appel vingt metre appel vingt metre, CQ vingt metre, ici doubla-vay neuf kah en ee (W9KNI) doubla-vay neuf kah en ee, appel vingt metre appel vingt metre..."

"W9KNI, W9KNI, ici Tay oo deux err koo, tay oo deux err koo qui vous appelle et passe a l'ecoute."

"Roger. TU2RQ, ici W9KNI qui revient, et merci beaucoup, cher ami..." Darn. A TU2; I have that one confirmed. But what the heck. You can't expect to call a sloppy CQ badly in someone else's language, and expect to only get new countries to respond. The fellow's doing me the favor in responding. I better pay attention here or I'm going to be in real trouble.

I follow my cheat-sheet closely, then turn it back to him. He comes back. Thank goodness, he seems to realize that I'm certainly not proficient in French; he gives my signal report very slowly, and then does his QTH and name letter by letter.

But getting Abidjan correctly would have been a bit tough if I hadn't read it off a QSL on the wall as he was spelling it out for me. His name, Claude, turned out to be easier, he didn't have to do it letter by letter for me. But he did.

What's that he's saying? He's going slowly and deliberately, and I think he's trying to be helpful; that would mean it has to be either about his rig or the weather. Yes! Emitteur means transmitter. Right. OK, 150 watts. And the French words for antenna and yagi are almost exactly the same as in English. Douze metres; yes, he's telling me that his yagi is up 12 meters; that's about 40 feet. Not bad. OK, he's turning it back.

Back to the cheat-sheet. I tell him about my rig and antenna, then the weather. I comment that my French is QSO French only and that I know very little more, so please excuse me, QRU etc. and turn it back to him again.

"Roger, W9KNI here's TU2RQ. Fine business, Bob. Not to worry, I speak English too. But you started in French, so I stayed with French. It is very nice to work an American station speaking in French, and your QSO French, as you call it, is very good..." He is, of course, being polite, my QSO French has to be very poor, and an assault upon his ears and his sensibilities, but he did understand it OK. And he seems genuinely pleased that I tried.

We ragchew a bit further. He tells me he is from the Loire valley in France, and that he is an economist and a civil servant of the government of the Ivory Coast. I tell him of the one time I visited Paris, and of my misadventures in finding a decent hotel. He finds that very funny; it seems he had an identical experience the first time he visited Paris.

I ask about stations in other parts of Africa; about the ever elusive TN8BK, about possible TJ1's and 5U7's. Claude tells me that TN8BK is actually fairly active, almost daily, but that he rarely signs his call, and only QSO's in French, though Claude is fairly sure that Dr. Bernard speaks fluent English. BUT, he is on holiday in France at present, and Claude does not think he is expected back until March.

However, Claude also has heard several times recently two new stations from there, TN8AJ, who apparently QSO's mostly in German and a little English, and TN8BP, who QSO's in French. So there yet may be hope. Claude tells me as well that there are at least four TJ1 stations on; one he says QSO's only in Dutch and mostly on long schedules to a girl friend back home; but the other three are active in French, and he is confident that if I watch for them in French, I am sure to get at least one of them.

He agrees that this a good time of year to chase them, because he can only work Europe in the evening hours on forty and seventy-five meters, and he doesn't like either band, as forty is full of very strong broadcast signals beamed to Africa by the different European countries, and the tropic QRN levels on seventy-five meters makes contacts unpleasant.

We end our QSO, having had a very pleasant rag chew. I have a very definite feeling he would not have come back to me if I had not tried the French CQ. There are a number of stations calling him when we sign, but he comes back to none of them, apparently having gone QRT. Well, I didn't get a new one out of it, but I had a very pleasant QSO, and got some excellent DX info as well. I'm certainly going to try it some more...

My further attempts at French CQ's over the next few days yield no spectacular results. I keep working at it, however. Each day as I drive to work, I play the alphabet tape over and over, which also contains a few simplified QSO's in French. And I listen a good deal on twenty meters in the French phone band. Most of the French speaking stations seem to hang out between 14105 and 14130, and I listen to them eagerly.

My continued practice improves my understanding to the extent that I can recognize most call signs with ease and accuracy, and follow most QSO's reasonably well, at least into the second or third "over." And I enter into QSO's in French more often as well. Most of the French speaking stations I QSO seem extremely pleased with my rudimentary efforts, even though I find out that almost to a man they speak far more English than I speak French.

I have one other very interesting QSO as well, when a YV5 answers one of my French CQ's. His English is rather poor, but his French seems excellent; at least it's a good deal better than mine. So we have a surprisingly good QSO in French. It is a second language for both of us, and probably that is why we do so well. But still, I do not raise any of the good Africans I seek.

Things start turning better the following weekend. I am listening carefully to the French phone band on another cold,

THE COMPLETE DX'ER 173

blustery Sunday afternoon, when I trip across a station signing clear with another station. The station transmitting is going through the usual courtesies, and signs the call signs; yes, "FY7SR ici TN8BP, Brazzaville, qui passe en QRT. Au revoir."

It's a TN8! Needless to say, I become like an octopus in a paper hanging contest, as I try a panic call on CW, since we W's have no phone privileges in that part of the band. But my luck in that sort of situation runs true to form - zero. He must have gone QRT as he had said he would.

But, all is not lost. I heard him, and now I have a time and frequency to watch. And, while I can't say I'm sure to recognize his voice again, at least I have some idea what it sounds like, which is sure to help in my hunt. Not to mention the very fact that I know now for sure he is on helps stiffen my resolve to keep watching for him...

The next weekend turns into a bust for me in spite of my desire to watch for the TN8. A neighbor asks for help Saturday to help move his sick mother; and he is a fellow who would drop anything if I asked him for help. Besides, he suffered TVI for two years without ever saying a word. Once I found out, accidentally, a high pass filter provided an instant cure. He didn't know it would be easy to fix, and didn't want to bother me. Needless to say, I help him move his mother. When I get home, the band is deader than King Tut's pet rock.

Sunday turned out no more productive from a DX'ing point of view, when a classmate of my wife's comes in from out of town with her husband. He turns out to be a delightful fellow, and we enjoy their visit - but no DX'ing for W9KNI...

The following weekend, I hope for better things. There is little I can do to prevent the kind of events that messed up last weekend, except hope for better luck. Saturday morning provides a good opening into Europe on fifteen, while twenty provides a nice long path into the Middle East. But nothing that I need shows up on those openings. As the morning goes by and the propagation paths shift, I start watching the French phone area on fifteen, around 21,200. A number of French stations are on, but most of them turn out to be VE2's in Quebec, plus several FM7's and an FG7. I find a lonely TR8 calling CQ, with a rather weak signal and no answers. But I don't need a TR8.

After lunch, I start working twenty meters over carefully. At first, I don't hear very much from Africa, but then I start to find more signals. A 5N2 operating higher in the band is S-7; I store his frequency, as he seems intent on a long session of working people for some special activity day, and I can use his signal as a beacon to check for signal strengths as the afternoon progresses.

The hunting gets easier as I go. The European signals fade out, leaving any signal in the French phone band likely to be either VE2's, FG's, FM's, FY's or Africans. I have to run out for about twenty minutes to pick up the kids at the library, which turns into a forty-five minute trip since they are not ready as promised, and I have to park and go look for them.

When I get back, I check first for the 5N2. Sure enough, he's still passing out QSO's, with a good following of stations pursuing him. And his signal is impressive now; well over S-9, a very good sign indeed for my cause. I flip back to 14,100 and start tuning through the first twenty-five kilohertz. I check the exciter and the linear; yes, they are ready, should I get lucky.

And there he is! "...soixante-treize, cher Jean-Paul, et j'espere que nous vous rencontrerons bientot. VE2BKT ici TN8BP, Brazzaville. Au revoir." I quickly set the transmit frequency so I have about a six hundred hertz note on the frequency and start calling the TN8, very slowly and deliberately, perhaps ten words per minute. The TN8's proficiency in CW is perhaps questionable, as there have been no reports at all of activity on CW by him in the DX'ers Tout Sheet.

I give a three by three call, more than I usually would, but I want to be sure to attract his attention. And, I do!

"Hallo the Whiskey Nine station, hallo the Whiskey Nine station on Morse, here is TN8BP. Old man, I regret I must QRT now, and I have not the time for the QSO. But, if you will call me tomorrow at this time, twenty one fifteen Greenwich time, on one four three zed zed, I will watch for your single sideband signal. For now, 73 and I QRT. This is TN8BP. Au revoir!"

I return, even more slowly, as it is obvious that his CW is a bit on the rusty side, "R R QSL QSL 73 73 DE W9KNI W9KNI W9KNI SK SK"

"Roger. W9KNI OK. Until tomorrow." And he is gone.

Wow! Already, I have a rudimentary QSO, but for sure I'll be waiting, loaded for bear tomorrow. I quickly make note of the time for the sked, then log this QSO. I'll certainly be there tomorrow...

Sunday, I am all nerves, having slept miserably the night before. It's been a long time since I've nailed a new one; and here I am with a schedule with one of the rarest. I'm more or less used to waiting for the DX'pedition to show, knowing I'm going to be fighting it out at the front lines with thousands of other eager DX'ers; that always keys me up, but nothing like this.

Finally, I trip across one of my G pals calling CQ, and I call him. He comes right back, and we have a very nice pleasant ragchew on QRQ CW. The challenge of keeping my trusty Bencher paddle dancing at forty words per minute takes up my attention, and I begin to settle down. But the band starts to fade to Europe in the early afternoon, and we sign clear. And I start to get nervous again.

It's 1900 Zulu. And I have better than two hours to go. I'm getting more anxious. My wife giggles at me; she is sympathetic, but she finds it amusing at the same time.

"Why don't you relax and go for a walk? It's at least twenty degrees out, and not very windy. The walk will do you good, and I'll have hot water for tea for you when you get home. Go on and do it."

She's right. A bit of exercise is what I need to settle down, and with two hours to go I need something. I put on my boots, my heavy coat, scarf, hat and gloves, and head out the door.

The sun is very bright, almost painfully so, as it shines off the snow that lies everywhere, especially bright compared to the soft indirect lighting of the shack. My eyes take a bit of time to adjust to the brightness. The air is cold. I walk rather briskly, and as I walk I feel the nerves easing up, and my stomach settling down. I'm almost able to laugh at myself; an adult man getting up tight over a few seconds of noise out of a box.

The walk turns out to be an ideal solution. In less than an hour, I'm back home, but much more relaxed. I sit in the living room, drinking tea and reading through the Sunday newspapers. After a bit I glance up at the clock. Right; it's 2:48, 2048 Zulu. I head downstairs into the shack, and start checking the gear out.

Lemesee here. I look around 14,300; not a part of the bands that I normally frequent. I find a clear frequency on 14289, and trim my tune up. Yes, it all appears to be functioning properly. I check the SWR; it is within reason. I check my Second Op to confirm that the bearing for TN8 from the American Midwest is 84 degrees. I start looking for a clear frequency around 14300.

It looks like that won't be easy. There seem to be QSO's all through the area. I glance at the clock; it shows 2058 Zulu, and I want to start calling. I don't really expect the TN8 to come back yet, but equally I certainly don't wish to leave any chance uncovered.

I get a bit lucky. I find a station running a phone patch at 14302, with what is obviously a long distance telephone call as one leg, which suggests the call won't last forever. And sure enough, within five minutes the two stations, a KL7 and a W3, end the patch and the QSO, and I jump all over the now clear frequency.

Before anyone else spots the clear frequency, I start my call. "Tay En Wheat Bay Pay, TN8BP, ici Doublavay Neuf Kah En Ee, W9KNI qui vous appelle et passe a la ecoute." I wait a few moments, with no response, and call again. No response. I call again. Nil. It's 2108. I know I'm early, but I want to be sure to have a frequency staked out. Somebody starts calling CQ about five hundred hertz lower. I listen; it's a WB4, saying that he is in Fort Lauderdale, and calling CQ. I turn the antenna down to the south-east, and when he breathes for a moment I inform him that the frequency is in use. He doesn't answer, but he doesn't call any more either.

Quickly, I swing the antenna back to an approximation of 84 degrees again, and call, again in French. I know that TN8BP speaks at least some English, but the last thing I need is to draw a crowd of stations waiting for him with me, and calling him in French is likely to stir few or no ripples. It's 2111 Zulu.

I call again, then listen. Oops, "K9AM, here's K6AAR, this spot looks clear Bob."

They don't have to wait long for a response from me. "I'm sorry, K6AAR; the frequency is in use. Here's W9KNI."

"Roger. Sorry about that. Bob, I'll go down another five and down from there. K6AAR, QSY."

I call the TN8 again, again in French. Suddenly, someone is calling me, not in French, but in English. English with an accent. My heart stops, my hair stands on end, my stomach starts a barrel-roll. "W9KNI W9KNI, here is TN8BP, TN8BP. Do you copy? Over."

"Roger, Roger. TN8BP, this is W9KNI. Roger, and thank you so much for meeting me here, my friend. I will give you a signal report on your next transmission. My name is Bob, Bravo - Oscar - Bravo, and my QTH is near Chicago, near Chicago. You are a new country for me, and I am very pleased and excited to contact you. How do you copy my transmission? TN8BP, here is W9KNI, over."

"Roger, my friend Bob. W9KNI, this is TN8BP. Very fine, Bob, and I am very pleased to meet you for this schedule. Your signals are good here, you are five by nine, five by nine here in Brazzaville, in the Republic of the Congo. My name is Emile, Emile. I am very pleased to be a new country for you, my friend Bob. Please send your QSL to F6TLK, F6TLK. He is my manager. Bob, I must go QRT now. The power here will go off very soon. Do you copy? W9KNI here's TN8BP."

"Roger, my friend Emile. TN8BP, this is W9KNI. Very fine Emile, and I understand about the power, so I shall not keep you. I will QSL via your manager, F6TLK, and thank you so much again for meeting the schedule. Very best 73's to you Emile. TN8BP, here's W9KNI signing clear. Au revoir!"

"Roger, Roger dear Bob. my best wishes to you and your family. W9KNI this is TN---."

And just like that he is gone. He wasn't kidding about losing power.

Gone, but not forgotten. Three or four stations suddenly start snarling over him. They are all W3's; someone's DX repeater must be running red hot. But Emile is gone, as I knew he would be. As I sit there smugly noting the details in the log book, I listen idly to the chatter on the frequency.

"Boy, did you hear that? That W9KNI had a TN8 come back. I didn't even know one was on, and then I hear one calling that W9. What do ya got to do?"

"Yeah, that's the truth. Some guys sure are lucky. OK, Charlie, thanks for calling that one in. I'm going back to watch some more of the Hula bowl."

I sit there, grinning from ear to ear over that exchange. Yup, I sure was lucky. Lucky that I went to the trouble to learn some French. Lucky that I happened to go digging through the French DX phone band every weekend for months. I start tuning down to 14100 again. Maybe I'll get lucky on a TJ1.

Notes: Since this chapter was written for the original edition, my friend W9BW, George Hanus, died tragically of cancer at a very premature age. This chapter is dedicated to his memory. Vale, George.

DX'ers interested in learning how to QSO in other languages would be well advised to first learn the basics of the subject language through traditional techniques. However, once having done so, they will find a book published by two Finnish amateurs, OH1BR and OH2BAD invaluable in learning to use that language for amateur radio. The book, spiral bound for convenient desktop use, is entitled "The Radio Amateur's Conversation Guide," and contains extensive QSO information in nine languages, with supplements of five other languages also available. Even without training in a language, the book teaches the rudiments of making QSO's in a foreign language.

In the United States the book is available from some amateur dealers, and from: CQ Magazine, 76 North Broadway, Hicksville, New York 11801.

Amateurs in other countries may wish to contact OH1BR directly: Jukka Heikinheimo, C/O Transelectro Oy, P. O. Box 118, SF-15111 Lahti, Finland, enclosing two IRCs or a dollar for an airmail quote and description of their offerings. Tapes demonstrating QSO's as printed in the book are also available, and are very much recommended for learning proper pronunciation, especially for the alphabet.

23
Winning, Losing and Playing the Game

Winning in the pile-ups is fun, a lot of fun. And losing is not fun, especially when an all-time new one is at stake. But one of the facts of life of each and every pile-up is that each time the DX station makes a QSO there are many losers as well as the one winner.

As we have seen in earlier chapters, when you are the loser, you try to learn from each loss; what are the habits of the DX operator, where should I transmit my next call, was my timing wrong or my call too long?

But sometimes, the experienced DX'er gets into a situation where it quickly becomes obvious that there is no hope of a QSO. And it is then the frustrated DX'er shows the world, or at least himself, what kind of person he is. Does he

QSY, or does he cause needless or even intentional QRM?

And sometimes, even though it would certainly seem that a QSO with a DX station is possible, or even probable, no matter what the DX'er does, he is wrong. He zigs; the DX station zags. And vice versa. Building character is all very well, but it's hardly as much fun as winning.

Yet, in a pile-up, by definition, for every winner there are losers, the operators whose calls are not being written into the DX station's log.

The DX'er is not alive, and never will be, who has gotten to DXCC by working each country on a single call. We are all losers at least a part of the time, and in truth most of the time. And that most certainly includes the author.

Losing is a part of DX'ing. If every DX'er always won on every call, attaining the Honor Roll would be no big deal. We all have to lose, to be losers, so that we can become winners too. Winning and losing are intertwined parts of DX'ing.

Do you like to lose? At times it actually can almost be fun; like when you are intentionally running five watts, chasing after one you don't need, and trying out new techniques of pile-up cracking. But for the committed DX'er, losing QSO after QSO in a pile-up over what would be a new one is a frustrating experience. And no fun at all.

First, a comforting thought. A fellow does not go to the trouble to get a license at a rare DX QTH, get some sort of station together, erect antennas, and print QSL's to be QRV for only a single operating session. There will invariably be another chance.

Now, let's turn the situation around a little. You are having a careful, patient tune of twenty, needing little, and not expecting to find a new one on, but you tune because you enjoy the thrill of listening to an open band. You trip across a fairly decent pile-up; naturally you punch the synch button and listen intently. Yes, there's the DX station transmitting; OK, it's 9K2DR.

You don't need Kuwait; you've had it on the wall for a long time. But you have noted that it seems to be rather rare of late; that would explain the rapidly growing pile-up. You listen carefully for a minute or so; it becomes quickly obvious that a lot of the people in this pile-up do need Kuwait.

What do you do now? Do you fire up the linear and go in after him? After all, it's been a while since the joy of the last big pile-up. And you've always gotten out well in that direction. But then you remember the anguish you suffered when you were chasing a 9K2 for your first Kuwait QSO. And you leave the linear off. You tune off frequency for a moment to a clear spot, and reduce your power to 10 watts output. You grin; now you are ready to go mix it up in the pile-up. You probably won't be successful, but it will be fun. And if you are successful, a real thrill.

One of the nicer aspects of DX'ing is that you set your own handicap. If you want to chase DX only on Sundays, fine. If you are independently wealthy, quit your job and move to the ideal QTH complete with stacked rhombics and a resident technician, great. If you want to be the first fellow to attain the Honor Roll running one watt output, good luck.

The DXCC and WAZ awards check QSL cards for compliance with the award requirements only. Any other aspect of the QSO, be it on the QSL or not, is ignored. The DXCC department of the ARRL will NOT check your cards against the calendar to certify that all QSO's were made on Sunday mornings local time. And no one will mark your QSL to show the power you used, be it 10 watts or 10 kilowatts.

It is equally obvious, then, that when you hang wall paper with stickers on it in your shack, you are the one who knows how the DX was worked. If it was done with one watt, you take great pride in the operating accomplishments, as indeed you deserve to. If working DX on Sunday only is your thing, you will know how well you have conformed with your goal.

And, if you did it with 10 KW, you will know that too. And even if you know that some other DX'ers run 10 KW as well, you must also be aware that others, a lot of others, run the power that their license allows. Taking a 10 dB advantage into a pile-up and beating out the horde is nice. But it isn't fair. And you know that too. You know you cheated. Even if no one else does. You set your own handicap.

24
DX'pedition!

My "DX'ers Tout Sheet" arrives, a day later than it should have, as usual. But, my unhappiness over the delay evaporates as I read the front page headline, "CLIPPERTON SOON." Clipperton! Always one of the rarest of the rare, one that has not been on the air for years. An uninhabited tropical atoll, French owned, surprisingly close to Mexico, and indeed, an island probably often visited by tuna fishermen.

But, getting a license and landing permission from the French government has appeared to be an impossibility for many years, as a DX'pedition to the island years ago had to be rescued at great expense to the French government. So, Clipperton has languished near the top of many DX'ers need lists for years. And it's one I need.

The story in the Tout Sheet gives all this information and more. It seems that a joint team of scientists and radio amateurs have formed a coalition to share expenses for an expedition to the island, and they have already secured all necessary paperwork! The government of French Oceania is requiring that the trip originate from Tahiti, not wishing to lend any credibility to Mexican claims to the island even though Mexico is a great deal closer. The call has already been issued; it is FOØCI, for Clipperton Island.

A sailing vessel, the "Tannu Hiva", has been chartered, and a crew of five French and American amateurs, along with three French scientists, are to make the trip. They are to be on the island for ten days.

And the departure date for the DX'pedition is only four weeks off, with estimated first operation ten days later. The DX'ers Tout Sheet obviously got a scoop on the story; there has been no talk on the bands beyond the occasional unfounded rumors.

According to DX tradition, standard operating procedure suggests I should schedule a two week vacation to Europe commencing on the date the DX'pedition has promised to first come on the air. That way, about the time I get back from my trip, they should be just getting ready to leave Tahiti for Clipperton. DX'peditions NEVER start on time, especially if it's one that you need.

I read further down into the story. Calls are listed; I look them over. I recognize two of them, but none of the rest. But the Tout Sheet tells a bit about the amateur members of the DX'pedition. Two of them are well known in contest circles, with one a Sweepstakes specialist. If he's a good Sweepstakes man, he should be ideal on a DX'pedition. Often a good DX'er goes on a DX'pedition and proves not very effective at the other end; it's not particularly surprising, since the skills needed to be a good DX hunter are far different from those needed to be the DX station working a huge pile-up in a fast and effective manner.

Ah well, it shouldn't make a lot of difference. With ten days time on the island, if all goes smoothly, the whole DX world should have an excellent chance to log the island. And the island is not too far from the equator, which means that even with poor conditions propagation should be at least adequate. And from the American Midwest, it should be an easy chip shot. I can't wait!

As the weeks go by, the bands and the DX bulletins are full of stories about the upcoming trip. The usual pleas for financial assistance go out. And, wonder of wonders, it begins to look like the group may actually leave within a day or two of the date promised.

Weekly, The DX'ers Tout Sheet carries the latest official information on the trip. With each issue, one set of problems are reported resolved, and new ones surface. One operator has to drop out of the DX'pedition, another is named to replace him. The captain of the "Tannu Hiva", the sailing schooner that is to take them to Clipperton, discovers that the group plans to take a number of fifty-five gallon drums of gasoline for the generators; his response is brief; "Non!" A compromise is reached; he will accept fifty-five gallon drums of diesel fuel, but no gasoline.

A mad search for portable diesel generators is conducted; and two are finally located in Ohio a week before the DX'pedition is scheduled to depart. An emergency appeal for assistance raises funds for air freight charges to ship the generators to Tahiti. A major argument ensues with the customs officials there, but finally a direct appeal to the governor-general helps arrange a nominal bond allowing the release of the generators, only forty-eight hours before embarkation. That the three scientists on the expedition are all French certainly proves to have been an excellent move on someone's part.

Finally the departure date arrives - and passes. The captain of the "Tannu Hiva" does not like the looks of a large low-pressure cell a thousand kilometers away, and suddenly twenty meter SSB is full of expert opinion on tropical depressions. There is also some mention of DX'er depression, as well. Then, forty hours later, and to everyone's surprise, the group departs on a favorable tide and wind.

For twelve days, the group wends its way cross the tropic sea, crossing the equator, onwards towards Clipperton. Nightly, a thousand DX'ers listen to 14,260, the official maritime mobile frequency of the DX'pedition. Tales of boredom, of sunburn, and of seemingly terminal mal-de-mer come from the "Tannu Hiva". Furtive phone-patches home are listened to on 14,340, where the American operators of the crew pretend to be maritime mobile on American flag vessels.

One of the DX'peditioners appears to be have the soul of a poet, and his nightly schedules with his buddy near Minneapolis are full of the wonder of porpoises and flying fish, of sharks and albatrosses, of night time skies jammed with stars in strange constellations and of mornings at sea bright and fresh. His enthusiastic descriptions weave a magic that captures the imagination of all that listen, and they are many.

Finally, early one Tuesday evening, the crew reports excitedly that they have made landfall, with Clipperton clearly in sight. A spontaneous cheer goes up on the frequency, as fascinated DX'ers applaud. Because of the failing light, the crew decides to land the following morning, and start a celebration aboard ship, celebrating the successful landfall.

Wednesday, I come down with an unfortunate illness, and find myself unable to report to work. And it IS true that for some reason I had slept miserably the night before. It is also true that I had told the boss about the situation, and warned him that I might well take the day off to play radio. After calling the office, I am able to get back to sleep for a couple hours. After all, Clipperton is an hour behind Central Standard Time, and there's no way that they can be QRV before 11 AM local time.

At 10:30, my wife wakens me, and trusts that the Clipperton flu will not turn out to be more than twenty four hours in duration. Needless to say, I heartily concur, even admitting my doubt that she is likely to catch the strange malady. After coffee and toast, I ease into the shack, and fire up the gear. A quick check of 14,025 and 14,195 yields no sign of any pile-up; a further check of 21,025 and 21,240 offers the same result.

I busy myself for a few minutes making sure all is in readiness. Yes, the antenna is on 216 degrees as close as I can hit it, the bearing my Second Op gives. All the gear checks out. I am afraid to turn the antenna to check the band out; I had a nightmare last night that I had turned it to Africa, and while beaming at 90 degrees the rotator had frozen. Why take the chance? Besides, propagation would have to be really bad not to have a path to Clipperton.

The bands seem rather quiet, except for what appears to be an inordinate number of strong carriers near the frequencies the DX'pedition has said it would use. At 21,255, I hear a station calling FOØCI, but it quickly becomes apparent he is calling blind, hoping to get lucky. When he finishes his call, perhaps thirty stations take the opportunity to describe their opinion of this novel tactic in terms some observers might call distinctly uncharitable. Within moments, the frequency quiets down, but not before I begin to get an idea that the Clipperton flu has reached epidemic proportions.

I spend the rest of the morning going from band to band, checking regularly all the posted frequencies, all without success. Ten meters seems to be dead, but fifteen and twenty are OK, and indeed both seem to be in decent shape. On fifteen there is a fairly healthy pile-up on an 8Q7 station, and I can copy the fellow weakly, even though I should be getting 30 dB attenuation in that direction with my antenna on Clipperton. But I don't need an 8Q7.

I sit there, listening carefully through the noon hour, with no result. My wife brings me a sandwich and a glass of milk around 1PM, 1900 Zulu. I eat with one hand on the dial.

At 1320 local time, I check 21,025, and someone is passing out high speed reports. Yes! It's him! He just signed his call. The transmitter is still on twenty meters, and I make a panic QSY. The signal is not very strong, but the pile-up is growing fast. I start calling up five, as the FOØ says, and when I check the action up five, I find out just how many are people are home sick. What a zoo!

I keep calling, as I try to discern the pattern he is working. Darn; some traffic cop is all over his frequency. Wait a moment. The cop is saying something. Like PIRATE, PIRATE. Hmmm. Wonder if he's right. It's easily checked. I run the antenna around to the south, towards the south-east, and the FOØ's signal picks up dramatically. I give one more call with the antenna SE, and nail him. I log it, but I already know I've been had; someone is having his fun and games. And within a few minutes, the station disappears, as it is more and more obvious that the station indeed is a joker.

The afternoon continues on, with no real news or hint of the real Clipperton gang. I don't know whether to worry or not, although of course I'm anxious. Some major DX'peditions have set everything up completely before they make the first QSO; others have gotten onto the beach with a vertical and a small generator and started making QSO's within thirty minutes of the first landing, while getting the rest of the gear

set up. And I have no idea how the Clipperton gang plans to start. And from the comments I hear, neither do a lot of other fellows who, like me, are waiting.

By late afternoon, though, I am beginning to have real concern, as there is no new information. Then, around 1730 Z, a W4 starts telling a buddy he has just seen the latest weather maps made from satellite data, and there appears to be a major disturbance in the Clipperton area. I go upstairs and ask my wife to watch the evening news, and to call me just before the weather forecast.

She does, and soon I am running back up the stairs to see if there is anything to the report. There is, there certainly is; a large, ugly looking pattern of swirling clouds, so definite that the weatherman points it out for its potential effect on the Mexican coastal area around Acapulco. He says nothing of Clipperton, of course, but needless to say, my worries are becoming more profound.

I sit most of the evening in the shack, waiting and hoping for good news, but there is no news, nothing except obviously unfounded rumor. I get to bed fairly early, around ten PM. After all, I did sleep badly last night, and certainly there's little hope of Clipperton coming on with darkness there, and likely very stormy.

I rise very early the next morning, and start tuning twenty for any news. Again, I find nothing new. I bring in the newspaper and have a look at the weather map. The depression has been upgraded to a tropical storm, with speculation noted in the paper that it may turn into a hurricane before it strikes the Mexican coast. The storm is named "Flora". I dress and go to work, rather depressed about the whole situation.

About 11 AM, I get a call from a buddy in California; there is no news. At three, I clean off the desk, and head home early, with the boss's blessing. I am on the rig as soon as I get home; but still no news. I catch the weather again. Flora appears to be well clear of Clipperton, and aiming at the Mexican coast.

I keep checking the bands through the evening without success. Conditions on twenty are quite good, but that is no joy tonight. I go to bed rather disturbed; although I don't really know any of the members of the DX'pedition beyond meeting two of them at Dayton once or twice. But, with all the hype

THE COMPLETE DX'ER 189

prior to the departure, and the heightened awareness because of the rarity of Clipperton, DX'ers everywhere are seriously concerned; the DX'pedition has become an institution, with its nightly progress reports so widely listened to, and with the wishes and hopes of DX'ers world wide riding on the "Tannu Hiva".

Thursday and Friday go on with more of the same. Flora apparently slammed into Mexico at just below hurricane strength, and did a great deal of damage, though more from flash flooding and mud slides than from wind velocity. But there is no word at all from the crew of the "Tannu Hiva".

Friday evening, a report is received from some hams in San Diego. Apparently, strictly by coincidence of course, a Navy Orion anti-submarine plane has made a training flight off the Pacific coast of Mexico, then swinging out to Clipperton as the corner of a triangle and heading back for home. The crew had made several low level passes over the island, and had seen extensive damage to the palm trees, and what appeared to recently deposited boxes, crates and drums in considerable disarray on the beach and in the water of the lagoon. K6AAR, the navigator on the flight, was quite sure there was no other sign of life.

Opinion on the bands generally feels that the news is more good than bad; with a strong likelihood that the crew had evacuated the island in the face of the oncoming storm, and had likely ridden it out hove to on the schooner somewhere well clear of the island. Should this be the case, there would be little doubt they would be trying to slog their way back to the island in the afterwash of Flora, and could well come on to the air at any minute.

Of course, hope springs eternal, which was probably first written by a DX'er. And, on the other hand, it seems strange that there has been no communication whatsoever from the boat, via amateur or commercial frequencies, when before they were QRV nightly. Regardless, it is certain that the original plans have failed. The bands are full of rumors but all seem to be without substantiation.

Saturday morning, I sleep in rather late. Apparently the tension of the last few days has taken its toll of me. But by 10:30, I am up and about, and by 11:00 I am at the rig. I check out 14,260 for the latest news. Nothing is new, but one of the San Diego gang suggests that the commander of the anti-submarine group at the naval air station, KD6BGD, was distinctly unsatisfied with some aspect of the recent Orion training flight down the coast, and has indicated they are going to do it again today. Perhaps by nightfall we'll have an update.

I start to tune twenty meters again; trying to get back into a more normal pattern of activity. I keep telling myself that no news is good news, and I start listening carefully to the normal activity of the band.

A check of fifteen shows it to be in fairly good condition. West European signals are still coming through, and there are several good Africans QRV as well. The antenna rotator hasn't frozen, and all the gear seems to be in good order. I run across my friend Ivan, F3AT, and I fill him in on all I know about the DX'pedition. In turn, he has some excellent information for me on a TJ1, knowing that to be one of the last Africans I need. When I sign clear, a couple other stations sign simple 73's as well; apparently there was something of a European audience on frequency, hungry for the latest news.

I have a light lunch, and check again on 14,260 to see if there is any word from the Navy. Apparently, the training flight is supposed to include night flying, so the departure time for the flight was moved to a later hour. However, there should be at least an hour of daylight left when the plane crosses Clipperton; coincidentally of course. In any case there won't be any word for several hours yet, and most probably none until the flight has returned unless an emergency situation is discovered.

I slip the transceiver back down the band to twenty CW, and start to look around again. Slowly, I start wending my way up the band. There seem to be a lot of Europeans coming through, though no really outstanding signals. I keep watch for Africans, but hear nothing from that continent. I cross 14,025, where the Clipperton boys were supposed to be operating, but hear nothing except for some high speed ragchewing. I listen for a few moments; apparently they are talking about Clipperton too, but with no new information. It depresses me; I tune on.

I keep going higher in the band, finding nothing significant, although conditions certainly seem favorable. The eastern European signals seem to be getting stronger; with LZ's and the YO's running S-8 and S-9; perhaps a goody from the Middle East will put in an appearance. I look at the dial; I am at 14,055. I keep moving up.

There's a CN2; a rare prefix and once a rare country, now the same as a CN8. Above him is an ISØ with an excellent signal; another that was once considered rather rare. I keep going higher; often you find a good bit of DX lurking high in the band away from the mainstream of the DX world.

There's a rather weak backscatter signal; I listen; hmmm. The way he's sending, that fellow sounds excited. Let's see, we're on 14,067. Nothing spectacular about that. Darn, there's a little QSB; I'm having trouble copying the fellow. OK, he turned it back, but he didn't give the other fellow's call sign. It was a W8, though. I got that part OK. I try to find the station he's working. Nothing. I listen extra carefully. Nothing at all. It seems unlikely that someone is working split frequency this high in the band; that's the sort of technique you find much lower in the band, like around 025.

On a hunch, I switch to my forty meter inverted "V". That I can't hear the DX station with my yagi on Europe makes me suspect that whoever it is is coming through on a different bearing; any signal from the north-east DX paths is not going to be so much stronger for a W8 that he can copy when I can't even detect it in W9 land. Besides, the W8's backscatter was rather weak; suggestive of a different path.

My hunch pays off; there IS a weak signal in there. Although the inverted "V" does not hear anywhere near as well on twenty as my yagi, on the other hand it is non-directional. The signal is too weak to copy on the "V" although detectable, which is better than the yagi pointed at Europe can do. My curiosity aroused, I start to swing the yagi around. At 135 degrees, I have a listen. Nothing, dead. I check the "V" again. Same signal, same strength, still transmitting. No, wait, he just turned it over to the W8 again.

The W8 is very weak on the yagi, and little better on the "V". I'm just going to have to sit this one out, and wait till the other station starts transmitting again. While I wait, I decide to swing the yagi around another 90 degrees to 225 degrees, and see what happens. I do, and the W8 get's somewhat stronger on the backscatter, though still not really good copy. But he still sounds excited. His fist is a bit choppy, and his spacing is terrible. Just in case, I punch the synch button to lock both VFO's together.

There, he turned it back, and now I am able to copy the

THE COMPLETE DX'ER   193

other station much better. He's at least 449, and still not signing any call. "...OK WILL CU BK HR IN 30 MINS BATTERIES SHUD BE GUD AT LST FEW MR HRS ES THEN WE ARE QRT TNX VY MUCH 73 DE FOØCI SK QRZ?"

FOØCI? The Clipperton crew? I don't know what to think of this bit at all; frankly I am rather skeptical. Still, the beam heading IS correct. But the frequency is all wrong. And batteries? No way. Still, as the Old Bard said, work them first and worry later. I call, a simple, "DE W9KNI K"

Whoever it is comes right back, "R W9KNI 599 BK".

I respond, "R TU 559 559 73 ES GL DE W9KNI SK".

I pick up the two meter microphone; "Hey, I don't think I believe it, but there's a station signing FOØCI on 14,067, that's 14,067; here's W9KNI."

I get a quick response, "Hello Bob. What's the deal? Here's K9MM."

"Hello, John. Well, the beam heading seems correct, and that's about all. The frequency's wrong. And the fellow is talking about his batteries not lasting very long. Still, if he was a pirate, I would think that he'd set up on 025, and pretend to be real. Anyhow, he's easy to work. He's working an N5 now."

"Yes, I hear him. OK, I guess I might as well work him. Thanks. Here's K9MM."

I listen to them. John gets him easily, and asks the operator what his QTH is. The FOØ comes right back, signs clear and calls QRZ without responding to John's question.

"I don't know, Bob. That sure is a strange one."

"Yer darn right. Let's check the heading as carefully as we can."

"OK. There, he's working a VE1...Well, it sure looks close to me."

"Yeah, me too. I wish he were stronger; then I could null him off the side of the antenna. But he's way too weak."

"Same here. I dunno. It's possible, at least."

"Yes, I agree. I think maybe we better phone a few of the gang. After all, 'Work them first and worry later,' you know."

"Right. OK, I'll call all the ones before W9NZM. You get NZM and the rest. Here's K9MM."

"Roger. Here's W9KNI."

John and I call all the members of the club that the club want-list shows as needing Clipperton. I have eighteen calls to make, and get the usual responses; a few no answers, a few that don't believe it to the point that they can't be bothered, a few out for the afternoon, but will leave a message etc. Still, about twenty members show up, between John's and my efforts, and all get through easily.

We have just had the last one through when the FOØ starts calling the W8 blind. There is no response. The FOØ calls again, and again no response. The FOØ seems to disappear. We wait, listening, talking about it on the two meter DX channel. No one knows quite what to make of it all; no one really believes that they worked Clipperton, but again, every one is hoping. In the mean time a couple of the non-believers report in on the channel.

There, the FOØ is calling the W8 again, and still no reply. The remaining fellows on frequency get through with ease. The FOØ calls the 8 again; I look at my log. It is definitely over a half hour since I QSO'd the station, whomever he is. His signal seems to be starting to get chirpy. Wait, there, the W8 is back.

The W8 is still too weak to copy well, but I get the idea he has been on the telephone making calls on behalf of the other station. The W8 makes a longish transmission, and the most of what he is saying seems to be reassuring notes. He turns it back to the FOØ.

"R FB ES CPY OK MUCH TNX JACK ES KNOW IT IS A RL RLF FOR OUR XYLS BT THIS IS ONLY SET OF GOOD BATTERIES ES NOT KNOW HOW LONG WE CN BE QRV BT IF WE CAN GET BOAT ENGINE GOING AGN WE BE QRV ON TRIP BK BT WE QSY TMW AS LOST ALL SUPPLIES IN STORM ES ONLY ENUF FOOD ES WATER FOR RETURN TRIP BT PSE QSP INFO TO GANG ON 14260 SSB ES WE PROMISE AGN NEXT YR BK QRU NW ES WE BE QRV TILL BATTERIES GONE THEN WE RETURN TO BOAT AND LV FOR HOME BK 73 ES TNX AGN VY MCH W8OWK DE FOØCI KN"

I start breathing again, stunned by it all. There is much left unsaid, much reading to do between the lines, but obviously a great deal has happened to the crew of the "Tannu Hiva" since they were last heard from. My belief in their being genuine is firm now. And it appears that in fact their batteries are nearly over the hill. The signal is definitely weaker, and the chirp more pronounced, obviously a sign of failing regulation as the batteries are dying.

There, they are transmitting again; "R FB JACK ES TNX AGN 73 QRZ DE FOØCI K"

There is a lot of chatter again on the two meter frequency, and several people, the earlier skeptics, are calling desperately now. But so are a lot of other stations in other parts of the country. I make a quick call to my buddy in California to tip him off; within seconds he is in the pile-up too, and scores on his second call. Our laggards make it through as well over the next six or seven QSO's, and it looks to be a darn good thing for them that they do; the FOØ signals are really getting weaker, and he is slowly moving up in frequency. The chirp is worse, as well. His batteries are clearly on their last legs.

At the same time, the pile-up is starting to get big, really big. The FOØ is asking for callers up five, and he starts having trouble picking calls out. He obviously is not using one of the regular rigs that was on the trip. And, suddenly, he starts dropping fast, and within 30 seconds he is gone. I sit there, staring at the rig, my imagination trying to fill in what has all the appearances of having been a desperate venture. And it looks like I have a legitimate QSO with Clipperton, as well!

The phone rings. I shake myself out of a daze. It is my friend Jim in California.

"Bob, did you hear the story that W8OWK is passing out on 14,265?"

"No, I can't copy him here, he's inside our skip zone. And besides, I'm still listening on 14,067. Tell me about it."

"OK, I thought that might be the case. Call me on 14,230, I've got most of the dope off the two meter net. See you."

"Right, thanks, Jim." I hang up; that new one minute telephone rate is super for quick schedules. I QSY to 14,230, and there's Jim calling me.

Jim passes along the story as learned from W8OWK. Though the FOØ had had to be brief, a good deal of information had been passed. It seems that the morning after landfall the crew had gotten all the gear ashore in good order, along with food, fuel, water etc. But the weather had started to deteriorate, and, instead of blowing over as they had expected, had turned into a major storm.

They had returned to the "Tannu Hiva" to ride out the storm. The "Tannu Hiva" had been well anchored in the lee of the island, but the wind had started to shift. They had tried to raise the anchor, but apparently it was jammed under some coral and they were unable to raise it in the face of the rising storm. Finally, they had had to cut the anchor line, as the wind shift was forcing them into dangerous waters. After they had cut the anchor, the winds had blown them far from the island. The captain and the native crew had managed to set a sea anchor, and the boat finally rode out the storm hove to, drifting slowly towards Mexico.

After the storm had abated, they had returned to the island, but had had to do so under sail, as the "Tannu Hiva's" auxiliary engine had apparently somehow been damaged while the boat was being tossed around, as had the boat's radio. That had ended any hope of communication, as the captain of the "Tannu Hiva" absolutely forbade use of the batteries for any purpose until the engine was started, as the batteries were necessary to turn the engine over.

When they had returned to Clipperton, they discovered that all the supplies and equipment left on shore had either totally disappeared, or had been rendered useless by the effects of storm and sea. And, that included the food and water planned for the shore stay.

The amateurs, the scientists and the captain of the boat had mutually discussed the situation, and it was agreed to stay on the island twenty-four hours, then begin the return trip to Papeete, Tahiti. All agreed to plan a return trip later, and the scientists felt that even a day's work would be very worth while, and make a return trip even more valuable.

One of the amateurs had kept a simple three watt direct conversion transceiver on the boat, with one set of lantern batteries to operate it, just "in case", and further had set up an emergency back-up schedule with his pal W8OWK if there were problems. Because of the lack of selectivity and the other limitations of the rig, a frequency high in the band had been agreed on.

And so it was that Clipperton was on the air. After the one brief operating session, nothing further was heard from the crew of the "Tannu Hiva" for nine days, when they made landfall at the Marquesas islands. There, they managed to get the boat's engine repaired and running, and hence were able to get the regular transceiver operational as well. And then the whole story was told.

The operating session had managed 53 QSO's from Clipperton, with 24 of them W9's in our club. We really were lucky on that one. And it certainly was exciting to hear our calls read out of the log as they completed the last leg of their trip, going from the Marquesas to Tahiti.

The operation had left one last question in my mind. What had happened to the second trip of the Orion airplane? I asked Jim about that. He found out, and I got the story from him later. Apparently, the Orion had had a hydraulic problem an hour out of San Diego, and had limped back home for minor repairs. The flight the next day had been canceled when word of the crew of the "Tannu Hiva" had been received.

It was weeks before regular QSO's seemed interesting again.

Author's note: This story, while fictional, in fact represents pieces of several true stories, and could easily have happened as described. Since the original publication of this story, Clipperton has become rather more available, thanks to changes in French government policy. But for me and for many long-time DX'ers, a QSO with Clipperton will always give a special thrill.

25
The Last Secret

There is one last secret vitally important to making the Honor Roll. If you have followed this book, or pursued a DX'ing career similar to that taught by this book, and if you are reasonably competitive, within two or three years you should have a score in the area of 275 to 290 countries.

But at this point your progress seems to come to a dead stop. You have cleaned the bands out; there is nothing on you need, or perhaps one exceptionally elusive station left for you to work. And you still are fifteen countries short of the bottom rung of the Honor Roll ladder.

You have developed good listening habits; you can wring the DX out of any band. You can almost always tell a DX station from a W, and you can almost always recognize a W

THE COMPLETE DX'ER

station in the act of QSO'ing a rare one. Your ears can knife through SSB QRM to copy the report from a weak and rare one right under the edge of a phone patch. You can copy lousy fists from weak stations on noisy bands in heavy QRM. You can predict secondary propagation paths with considerable accuracy to most any part of the world, and that includes the long paths.

In short, you are a complete DX'er.

But, after a couple of years of unremitting and unrewarded tuning, the thrill of it all begins to pall a bit. Sure, some DX'pedition will be along sooner or later to add to your total. And you know you will be able to work him with little trouble if everything goes well. And, the pile-ups will be fun, besides.

However, DX'peditions to the hard to reach places occur only on a rare basis. And after a couple more years, you have confirmed everything except the places that are either hard to reach or impossible to get appropriate permits and licenses for. And you are still seven countries from the Honor Roll.

New ones come very slowly; two a year at this point makes a very good DX year. Boredom sets in, and too often the DX'er gives up DX'ing and takes up other interests. After all, to have become a good DX'er took a lot of time, and required the development of multiple skills, honing the operator and his station into a highly efficient operating machine. If the skills start to lapse from lack of use, and the operator loses interest out of boredom, when the time comes to work the new one that finally becomes active, too often the terror of the bands a few years back is missing from the logs, another victim of malaise.

The secret to attaining the Honor Roll, then, is to cultivate new interests within amateur radio that are consistent with the station and antennas a DX'er needs. For example, take up QRQ DX ragchewing. In years past, to spend much time ragchewing was time wasted that should have been spent tuning for the new one. Now, there is no new one, and the ragchewing is the interest that keeps you QRV waiting for that new one. And, the overseas friendships formed can be tremendously satisfying, as well, not to mention the source of much good DX info on the last few that you need.

Certainly another important interest is contesting, and a lot of DX'ers spend considerable time preparing for the next contest. There is one potential hazard in becoming a serious contest operator, and that is that antennas good for some certain aspect of contesting may not be good for all DX'ing needs. An antenna that blew down two weeks after it did its job in the big contest is of no use three weeks later when the 3Y comes on. And some contesters, recognizing this, keep good tribanders always operational for just that need, not to mention their use in a contest to pick off multipliers.

Some DX'ers chase countries all over again using QRP with five watts or less RF out, and try to build significant scores that way. It certainly is an ideal way to wait for the new ones, as it helps keep your operating skills sharp and keeps you optimizing other aspects of your station. Just don't sell the big rig!

Another operating activity is hunting certificates and awards other than the DXCC and WAZ variety. At times it seems like there are an unlimited number of awards, and most DX'ers who are chasing them specialize in two or three maximum. Some fellows chase the Soviet oblasts; others work towards the tough Japanese Cities Award, the French DUF certificate, or RSGB's Islands On The Air. You can chase Swiss Cantons, Japanese prefectures, New Zealand counties, German states, or Bermudian parishes, to name a few. There are plenty of awards to chase, and there are good books available listing them.

You can check into traffic nets, you can keep schedules with maritime mobile stations, you can explore new modes of operating. The important thing is to keep your operating interest high, to keep the station operational, your skills sharp, and your juices flowing. That is the final secret of making the Honor Roll, both onto the bottom rung, and then to the top.

As a matter of interest, in the period since 1968, it has been possible to go from zero countries to the Honor Roll in never more than eight years, usually less. And this is on a reasonable basis, which is to say working every country that was at least occasionally available for all to work. Countries that have not been on the air for years do come back on, sometimes when no one would guess that there was any chance of operation.

Stay active, be ready, and you are sure to be on the Honor Roll if you want it badly enough. Good DX'ing!

26 Conclusion

And so it ends. Hopefully, you will have learned at least a little about DX'ing, and perhaps a great deal. And, I sincerely hope that at the same time you will have been entertained by some of the stories of life in the trenches, and perhaps learned therein as well.

By now you will have figured out that I love DX'ing. But the game is nearly over for me. At the present time, I need one country all time, Albania, and four on CW. This is not bragging, but rather regret that there is little more to do in chasing countries. But it was fun; a lot of fun, and it has changed my life as well, both career wise and in the wonderful friends I have gained, both world-wide and closer to home. Many of their calls are sprinkled through the book.

I have found other interests in high frequency operating, and they too are rewarding. I plan to be on the bands for a long time yet.

There are two points that I want to leave with you. One is that while I have tried to cover every significant operating technique used in the DX game, there are undoubtedly more, including some I never learned, and perhaps a couple I've forgotten. And there are a few I think very little of, and will not dignify by printing them. But in any case, you will have learned by now that every successful technique has a logical basis and a methodology, one that makes sense when thought out.

Perhaps you will think out a few new ones; but in any case whenever you hear anyone working a new one, ask yourself, "Why did that station get that contact? How did he do it? Did I not do something that could have helped me get the QSO?" That kind of analysis will help you keep learning new techniques, be they in the pile-up or in open band tuning.

The other point is to remind you that DX'ing in terms of the DXCC and the Honor Roll is a never ending game. That means, among other things, that if you don't make it through a pile-up for a new one you need, don't despair. You will get another chance, sooner or later. Probably sooner. For example, there is no country I have worked that I have not heard several times at a minimum, times that I could or should have worked the station. Be patient; that is the essence of the successful DX'er.

DX'ing is a wonderful hobby. Be responsible. Help make it better, not worse. If you get angry at what is happening in a pile-up, don't say what you think on the air. Observe the terms of your license. Support your national amateur organization, and be involved in it.

And thank you for reading this far; it's been fun writing it. Good DX'ing!

de W9KNI

THE COMPLETE DX'ER

Appendix

There are several references in the book to the "DX'ers Tout Sheet," a mythical weekly DX bulletin. While the Tout Sheet is fictional, there are a number of DX bulletins published which are most helpful to the serious DX'er.

The following list of bulletins is, undoubtedly, incomplete; it is a collection of English language bulletins that the author believes are available to any DX'er willing to subscribe. There are also some excellent DX bulletins published in other languages, and a note, with an SASE please, to the DX editor of one of the national society magazines published in that language will often bring a listing of several.

Obviously, there are a number of bulletins published. Each has a different perspective, and the DX'er should examine all of the pertinent ones before choosing. A note to the publisher along with a self addressed stamped envelope good for three ounces of postage will bring back sample back issues along with a subscription rate schedule, and probably notes on what the magazine is trying to accomplish.

All of the following magazines operate on a low budget; be kind and supply the return postage for your inquiry. And, having taken something from them, be sure to put something back in - send news to the editor. That is the only way that the bulletins can keep current on rare DX sightings.

Here then is the list, believed to be accurate at the present time, and presented in alphabetic order...

DX NEWS Sheet
C/O The Radio Society of Great Britain
Lambda House, Cranborne Road
Potters Bar, Herts EN6 3JW
England

INSIDE DX
436 North Geneva Street
Ithaca, New York 14850
U.S.A.

LONG SKIP
Box 717 Station Q
Toronto, Ontario M4T 2N7
Canada

QRZ DX
P. O. Box 832205
Richardson, Texas 75083
U. S. A.

THE DX BULLETIN
P. O. Box 50
Fulton, CA 95439
U. S. A.

THE LONG ISLAND DX BULLETIN
Box 173
Huntington, New York 17743
U. S. A.

Technical Notes

For those interested in such things, here are a few technical notes about the creation of this book. The initial manuscript was written in 1982 and 1983 on a Radio Shack Model 3 computer using tape based Scripsit. The tape files were later transferred to MS-DOS files and installed on a Tandy 1000 computer.

The present edition was rewritten on a Tandy 3000HL computer, using XyWrite 3+ as a word processor. Typesetting was done on a Hewlett-Packard LaserJet series II printer. The typeface used was Bitstream "Dutch", their version of Times-Roman. Lodestar XPT was used to integrate the Bitstream fonts into XyWrite for printing. The text was initially set in 18 point type, then photo reduced for greater clarity.

Widow and orphan controls, although available, were purposely omitted due to space considerations.